The Bio-Revolution

CORNUCOPIA OR PANDORA'S BOX?

Edited by Peter Wheale and Ruth McNally

PLUTO  PRESS

London • Winchester, Mass

Dedicated to the welfare of animals and the environment

First published 1990 by Pluto Press
345 Archway Road, London N6 5AA
and 8 Winchester Place, Winchester
MA 01890, USA

British Library Cataloguing in Publication Data
The bio-revolution: cornucopia or pandora's box?
1. Genetic engineering. Social aspects
I. Wheale, Peter, 1944- II. McNally, Ruth M. 1958-
306'.45

ISBN 0-7453-0337-4 hb
ISBN 0-7453-0338-2 pb

Library of Congress Cataloging in Publication Data
The Bio-revolution : cornucopia or Pandora's box? / edited by Peter
Wheale and Ruth McNally.
    p. cm. – (Pluto studies in biotechnology)
    "Based upon the eighteen papers presented at the Athene Trust
International Conference on Genetic Engineering which was held in
London in October 1988"–Pref.
    Includes bibliographical references.
    ISBN 0-7453-0337-4 – ISBN 0-7453-0338-2 (pbk.)
    1. Genetic engineering–Congresses. 2. Biotechnology–Congresses.
I. Wheale, Peter. II. McNally, Ruth M. III. Athene Trust
International Conference on Genetic Engineering (1988 : London,
England) IV. Athene Trust. V. Series.
TP248.6.B53 1990
660'.65–dc20   89-26460
                                                            CIP

Typeset by Stanford Desktop Publishing Services, Milton Keynes
Printed in Great Britain by Billing and Sons Ltd, Worcester

The Bio-Revolution

Genetic Engineering Series from Pluto Press

Series Editors: Dr Peter Wheale and Ruth McNally

Forthcoming:
The Gene Consumers' Guide
Genetic Engineering and Reproduction: A Family Guide

Contents

Part II: Genetically Engineered Bovine Somatotropin (BST)

Part V: Public Acceptability and Control of Genetic Engineering

List of Abbreviations

ACDP	Advisory Committee on Dangerous Pathogens (HSE, UK)
ACGM	Advisory Committee on Genetic Manipulation (HSE, UK)
AcNPV	*Autographa californica* nuclear polyhedrosis virus
AFRC	Agriculture and Food Research Council (UK)
AGM	Annual General Meeting
AGS	Advanced Genetic Sciences
AIDS	Acquired immune deficiency syndrome
APC	Animal Procedures Committee
APHIS	Animal and Plant Health Inspection Service (of the USDA)
A (SP) Act	Animals (Scientific Procedures) Act (1986) (UK)
AWA	Animal Welfare Act (USA)
BBRG	Biotechnology Business Research Group
BGH	bovine growth hormone
BST	bovine somatotropin/somatotrophin
B. thuringiensis	*Bacillus thuringiensis*
CAMR	Centre for Applied Microbiological Research (MOD, UK)
CAP	Common Agricultural Policy (EC)
CBI	Confederation of British Industry
CEC	Commission of the European Community
CERB	Cambridge Experimental Review Board (USA)
CIWF	Compassion in World Farming (UK)
cm^2	square centimetre
CUBE	Concertation Unit for Biotechnology in Europe (EC)
CVMP	Committee for Veterinary Medicinal Products (EC)
DES	Department of Education and Science (UK)
DHSS	Department of Health and Social Security (UK) (now split into Department of Health [DoH] and Department of Social Services [DSS])
DNA	deoxyribonucleic acid
rDNA	recombinant DNA
DOE	Department of Employment (UK)
DoE	Department of the Environment (UK)

DoH	Department of Health (UK)
DPAG	Dangerous Pathogens Advisory Group (DoH, UK)
DSS	Department of Social Services (UK)
DTI	Department of Trade and Industry (UK)
EBCG	European Biotechnology Coordinating Group
EC	European Community
ECLAIR	European Collaborative Linkage of Agriculture and Industry through Research (EC)
ECU	European currency unit
EDF	Environmental Defense Fund (USA)
EEC	European Economic Community
EPA	Environmental Protection Agency (USA)
EPC	European Patent Convention
EPO	European Patent Office
E. coli	*Escherichia coli*
FAO	Food and Agriculture Organisation (UN)
FAWC	Farm Animal Welfare Council (MAFF, UK)
FDA	Food and Drug Administration (USA)
FDCA	Food, Drug and Cosmetic Act (USA)
FIFRA	Federal Insecticide, Fungicide and Rodenticide Act (USA)
FLAIR	Food-Linked Agro-Industrial Research (EC)
FoE	Friends of the Earth
GILSP	Good Industrial Large Scale Practice (OECD)
GLSP	Good Large-Scale Practice (ACGM/HSE)
GMAG	Genetic Manipulation Advisory Group (DES, UK)
HGH	human growth hormone
HIV	human immunodeficiency virus
HMSO	Her Majesty's Stationery Office (UK)
HR	House of Representatives (USA)
HSC	Health and Safety Commission (DOE, UK)
HSE	Health and Safety Executive (DOE, UK)
HSW Act	Health and Safety at Work etc. Act (1974) (UK)
HSUS	Humane Society of the United States
IBA	Industrial Biotechnology Association (USA)
IBR	infectious bovine rhinotracheitis
IGAP	Institute for Grassland and Animal Production (AFRC, UK)
IGF	insulin-like growth factor
IISC	Intentional Introduction Sub-Committee (formerly PRSC) (ACGM, UK)
IoV	Institute of Virology (UK)
IRDAC	Industry Research and Development Advisory Committee (EC)
LFC	London Food Commission

MAFF	Ministry of Agriculture, Fisheries & Food (UK)
MEP	Member of the European Parliament
mg	milligram
MMB	Milk Marketing Board (UK)
MOD	Ministry of Defence (UK)
MOET	multiple ovulation with embryo transfer
NAVS	National Anti-Vivisection Society (UK)
NCC	Nature Conservancy Council (UK)
NEPA	National Environmental Policy Act (USA)
NERC	Natural Environment Research Council (UK)
NFU	National Farmers' Union (UK)
NIABY	not in anybody's backyard
NIH	National Institutes of Health (USA)
NIMBY	not in my backyard
NOAH	National Office of Animal Health
NPV	nuclear polyhedrosis virus
OECD	Organisation for Economic Co-operation and Development
OSTP	Office of Science and Technology Policy (USA)
OTA	Office of Technology Assessment (USA)
PAHO	Pan American Health Organisation
PIB	polyhedrin inclusion body
PFU	plaque forming unit
PST	porcine somatotropin/somatotrophin
PRSC	Planned Release Sub-Committee (renamed IISC) (ACGM, UK)
PTO	Patent and Trademark Office (USA)
RAC	Recombinant DNA Advisory Committee (NIH, USA)
RCEP	Royal Commission on Environmental Pollution
RSGB	Research Surveys of Great Britain Ltd
RSPCA	Royal Society for the Prevention of Cruelty to Animals
S	Senate (USA)
SIV	simian immunodeficiency virus
S. exigua	*Spodoptera exigua* (small mottled willow moth)
S. frugiperda	*Spodoptera frugiperda* (fall armyworm)
T. ni	*Trichoplusia ni* (cabbage looper moth)
TPA	tissue plasminogen activator
TSCA	Toxic Substances Control Act (USA)
TUC	Trades Union Congress (UK)
UN	United Nations
USDA	United States Department of Agriculture
VPC	Veterinary Products Committee (UK)
WHO	World Health Organisation

List of Figures

List of Tables

List of Boxes

Preface

Will the revolution in the biological sciences produce a cornucopia or open a Pandora's box? It was to address this important question that, in September 1988, the Athene Trust gathered together genetic engineers, veterinarians, animal welfare campaigners, ecologists, regulators, philosophers, theologians, farmers and industrialists at an international conference in London.

The Bio-Revolution: Cornucopia or Pandora's Box? is based on the proceedings of the Athene Trust Conference. The informed opinions of the contributors, each an expert in his or her own field, engage the reader in an exciting and serious debate over the potential benefits and dangers of genetic engineering.

Clear and concise introductory chapters, written by the editors, Dr Peter Wheale and Ruth McNally, guide the reader through each of the parts of the book, introducing the contributors, and contrasting and comparing their main themes and arguments on such controversial issues as: the creation of transgenic animals; administering genetically engineered bovine somatotropin (BST) to dairy cows; the deliberate release of genetically engineered organisms into the environment; projects to sequence and map the entire human genome; the patenting of life-forms, and the international regulation and control of genetic engineering.

Each of the five parts of the book is fully referenced and concludes with an edited transcript of the respective open forum discussions which took place at the end of each session of the conference. The editors also provide a comprehensive glossary of technical terms, a list of abbreviations, name and subject indexes and biographical notes on the contributors.

The aim of this book is to increase public awareness of the subject and to stimulate public participation in the debate over genetic engineering. No specialist knowledge is required to understand the ideas, arguments and discussions presented. It is intended for students of the life sciences, the social sciences and the humanities, and should prove useful to everyone concerned about animal welfare and the environment.

The Athene Trust is dedicated to the promotion of harmony between people and the natural world. The conference was organised by Joyce D'Silva, Education Director of the Athene Trust, and inspired by suggestions and ideas from Dr Michael Fox and Jon Wynne-Tyson. It was financially supported by Compassion in World Farming, the Royal Society for the Prevention of Cruelty to Animals (RSPCA), the Farm and Food Society, the Theosophical Order of Service, the Catholic Study Circle for Animal Welfare, Education Services and the European Commission's Concertation Unit for Biotechnology in Europe (CUBE).

The editors wish to express their gratitude to the Athene Trust for having provided them with the opportunity to edit these conference proceedings and arrange for their publication as a book in this series on genetic engineering for Pluto Press.

Dr Peter Wheale and Ruth McNally
London, February 1990

Part I
Transgenic Manipulation, Patenting and Animal Welfare

1 Introduction

Dr PETER WHEALE and RUTH McNALLY

Genetic Engineering

Recent years have witnessed what David Weatherall, Nuffield Professor of Clinical Medicine at Oxford, has described as a revolution in the biological sciences comparable to that in physics earlier this century (Weatherall 1987). Indeed, it was several eminent physicists, most notably Niels Bohr, the Danish nuclear physicist, and Erwin Schroedinger, the Austrian theoretician of quantum mechanics, speculating in the 1930s on the physical basis of heredity, who provided early inspiration for this 'bio-revolution'. Advances on both sides of the Atlantic in the theory of heredity and the study óf molecular biology led in the early 1950s to the elucidation of how it was possible for a molecule – deoxyribonucleic acid (DNA) – to convey hereditary information. This discovery was the foundation of molecular genetics, a science of enormous potential for the manipulation of the hereditary material of living organisms. This potential was realised in the early 1970s when the technology popularly known as genetic engineering was developed.

Genetic engineering manipulates genetic material by intervention at the cellular and molecular levels rather than at the level of whole animals and plants. This *micro*genetic engineering has made 'transgenic manipulation' possible – totally unrelated species that cannot interbreed can now be engineered to share each other's genetic material. By comparison with traditional breeding practices, *micro*genetic engineering reduces the time needed to produce new varieties of bacteria, plants, animals and viruses. More organisms of novel genetic composition are now being produced than would ever have been possible using conventional breeding practices or through the natural processes of evolution.

Genetic engineers have created viruses which are hybrids of two different viruses, and super-toxic bacteria for use as living pesticides. Bacteria have been genetically engineered for use as biological 'anti-freeze' to prevent frost-damage to crops, and toxin genes from bacteria have been inserted into plants to make them poisonous to insect pests. Genetically identical clones of animals and

3

plants have been created. Even human genetic material has been used in transgenic manipulation, so that there now exist sheep which produce human proteins in their milk, and pigs which contain human growth hormone genes.

The ease with which genetic engineers manipulate DNA on the laboratory bench belies the complexity of the systems which DNA enters into inside living cells and microbes. The first reports of 'jumping genes', deriving from Barbara McClintock's pre-molecular analysis of maize genetics, were mostly ignored until years later when they were confirmed by molecular genetics studies on microbes (Fedoroff 1984; Keller 1983; McClintock 1978, 1984). 'Jumping genes', or mobile genetic elements, of various categories, are now considered a major feature of all DNA, where they enable organisms to adapt to challenge at the genetic level through gene duplication and rearrangement (Shapiro 1983; Tonegawa *et al.* 1978; Watson *et al.* 1983; Berg and Howe 1989). The DNA of cells and microbes is now believed to be in a constant state of flux, caused through its rearrangement and duplication and the accommodation of DNA from external sources, in the form of invading viruses and plasmids (Day 1982; Scott 1987; Wheale and McNally 1988a, Chapter 4).

Transgenic manipulation depends upon the dynamic nature of cells and microbes, in particular their capacity to integrate pieces of DNA from external sources, duplicate them and express their messages as gene products. Genetic engineers often exploit the intimate relationship between cells and microbes and their viruses and plasmids to assist them in transgenic manipulation; they use viruses and plasmids to make 'vectors' to carry genes into the cells and microbes which they wish to manipulate (see Watson *et al.* 1983; Wheale and McNally 1988a, Chapter 2).

The dynamic nature of genomes and their intimate relationship with viruses and plasmids have important implications for the risks of genetic engineering. For example, the integration of the DNA introduced by transgenic manipulation could trigger the mobilisation, duplication and mutation of parts of the genome (Singer 1983; Fedoroff 1984). It is known that the mobilisation, mutation and duplication of DNA are implicated in the transformation of normal cell lines into cancerous ones, and are a source of the genetic variation which gives rise to genetic differences between individuals, morphological adaptation and abnormalities, and is the starting point for the evolution of new genes and the origin of species (Leder *et al.* 1983; Rose and Doolittle 1983; Scott 1987). There is also the possibility that the viruses and plasmids used as vectors in transgenic manipulation will become mobilised and escape from the cell or microbe, carrying the gene insert to other cells or organisms (see Lichtenstein 1987; Wheale and McNally 1988a).

The perceived novelty of genetic engineering techniques and of the organisms they produce has rendered genetically engineered living organisms and their parts and processes patentable. The first patented mammal was a transgenic mouse – the so-called 'cancer mouse' – which was genetically engineered using human cancer genes so that it develops cancerous tumours (see Box 1.1). By the end of 1988 about 50 applications for animal patents had been filed with the US PTO. The rights conferred on the holder of the patent on the 'cancer mouse' extend to acts of propagation for commercial use – the patent covers not just the mouse, but its lineage for 17 years. The simplest way of enforcing the monopoly of patenting rights over reproduction will be to engineer the animals so that they are effectively sterile in a similar way to many varieties of hybrid seeds.

Technical Merit

In this part of the book, the contributors analyse the potential benefits and animal welfare concerns arising from the transgenic manipulation and patenting of farm and laboratory animals. In Chapter 2 Dr Caroline Murphy, a geneticist who works as the Ethics Education Officer for the RSPCA, explains the basics of molecular genetics and transgenic manipulation. The problems encountered when scientists working in agriculture and animal breeding attempted to apply transgenic manipulation techniques to farm animals are discussed in Chapter 3 by Dr Grahame Bulfield. Bulfield is Head of the Edinburgh Research Station of the Agriculture and Food Research Council (AFRC) Institute of Animal Physiology and Genetics Research, which produced the world-famous transgenic sheep which lactate Factor IX, the human blood-clotting agent which is used to treat certain haemophiliacs.

Bulfield and Murphy describe three ways in which the transgenic manipulation of animals could potentially provide benefits to society. First, in 'molecular farming', where it is hoped that transgenic animals will be able to be used as 'bioreactors', like the Factor IX sheep, which secrete human pharmaceuticals in their milk. Secondly, transgenic animals could be used to investigate the control of gene expression and the regulation of animal development, and as models to study human diseases. And thirdly, transgenic manipulation could be used to produce commercially superior livestock – to make them grow faster, leaner or larger, and to produce more milk, eggs or offspring.

Box 1.1:
The Patenting of Genetically Engineered Organisms

The bio-revolution is responsible for commercial pressure to permit the patenting of living animals, plants and microbes and their products and processes. In 1980 the US Supreme Court granted a patent to Ananda Chakrabarty for genetically modified bacteria. The significance of this decision is that it established that, under US patent law, patentability of a product is not dependent upon its being non-living: a living product can be patented provided it can be established that it is artificially produced rather than a product of nature. Since the Chakrabarty precedent was established, many patents on living microbes have been granted by the US Patent and Trademark Office (PTO), the patenting offices of the UK, West Germany and Japan and by the European Patent Office (EPO) (see Wheale and McNally 1986).

The US PTO also led the way in the patenting of animals. It issued the first patent on a modified animal in April 1987 for a special breed of Pacific oyster. In April 1988 it issued its first patent for a mammal – a genetically engineered mouse, popularly known as the 'cancer mouse' or 'oncomouse'. Human cancer genes (oncogenes) inserted into the genetic material of 'cancer mice' guarantee that they develop breast cancer in 90 days. The patentors claim that the 'cancer mouse' can be used as a model system with which to study how breast cancer develops and to test new drugs and therapies that might be useful for the treatment of cancer in humans. The 'cancer mouse' was patented by Harvard University, which has arranged that Du Pont should have exclusive rights over the US patent. The patent gives the patent-holder proprietary rights over the offspring of the patented mouse and their offspring for 17 years. It also covers any mammal, excluding humans, that has had cancer genes inserted into its genetic material (genome) at the embryonic stage.

An application to the EPO to patent the 'cancer mouse' was rejected in June 1989 because the European Patent Convention (EPC) does not permit the patenting of animal and plant varieties. However, the patent attorney for Harvard University is planning to appeal against this decision, claiming that the 'cancer mouse' is not an animal 'variety' and is therefore patentable under the EPC (Dickman 1989).

Not all countries in the European Community (EC) are signatories to the EPC. In October 1988 the EC Commission published a

---

Box 1.1 continued ...

Proposal for a Council Directive on the Legal Protection of Biotechnological Inventions (European Commission 1988). If the proposed directive is adopted by the European Council of Ministers, it will become EC law and thus binding on each of the Member States. The proposed directive states that micro-organisms and any biological classifications of plants and animals other than varieties are patentable subject matter. It also states that parts of animals, plants and micro-organisms are patentable and that living matter which formed part of a pre-existing natural material is patentable. Unlike the US PTO ruling in the case of the genetically engineered mouse the European Commission's proposed directive does not specifically exclude human material and thus, under the proposed directive, human beings and their parts appear to be patentable subject matter.

In Chapter 4 John Webster, Professor of Animal Husbandry at Bristol University, analyses the various ways in which genetic engineering could be applied to improve the productivity of farm animals. His analysis covers the use of genetic engineering to improve disease resistance, to manipulate animal size, shape, composition and reproductive efficiency, to increase productivity through the use of genetically engineered growth hormones, and to broaden the range of animal produce to include pharmaceuticals. He concludes that the scope for producing useful changes in farm animals through the use of genetic engineering is much less than both its advocates and detractors believe.

Animal Welfare

Each of the contributors in this part of the book addresses the potential risk that transgenic manipulation and patentability pose to the welfare of animals.

They all agree that considerable animal welfare problems have been created through generations of traditional selective breeding. Bulfield claims that, by comparison with conventional breeding practices, the new techniques of genetic engineering are precise and controlled and that overall they should have a beneficial effect on animal welfare. This contrasts with the view of Dr Michael Fox, Vice-President of the Humane Society of the United States (HSUS), who argues in Chapter 5 that transgenic manipulation will increase animal welfare problems. Fox describes the adverse effects that transgenic experiments have already had on the welfare of animal subjects and identifies new welfare problems that could arise in the future as a result of transgenic manipulation. He argues that the new potential for patenting genetically engineered animals and their products provides an economic incentive for their increased genetic manipulation and exploitation. This is because patenting is regarded as a stimulus to research and development since it rewards companies, institutions and individuals who undertake such activities by granting them a limited monopoly over their patentable inventions. He reports that a coalition of animal welfare organisations, environmental and public interest groups, farm groups and religious leaders in the USA is asking the US Congress for a moratorium on animal patenting.

Fox believes that our moral duty towards farm and laboratory animals goes further than to refrain from inflicting unnecessary suffering upon them. He believes that the transgenic manipulation of animals is ethically questionable because it violates their *telos*, or intrinsic nature, and that it should only be undertaken for exceptional reasons, for example, to produce life-saving pharmaceuticals. Making a distinction between suffering caused through transgenic manipulation itself and the way in which manipulated animals are treated subsequently, Webster and Fox suggest that economically viable transgenic animals which lactate valuable proteins, such as the Factor IX-secreting sheep, are liable to be treated better than their non-manipulated counterparts.

Fox and Murphy draw attention to the potential for ecological damage arising from the deliberate release or escape of transgenic animals. They express concern that new breeds of transgenic animals may be able to colonise and destroy the few remaining wildernesses. There is also the risk that the foreign genes used in transgenic manipulation could be transferred to other animals, altering their properties with unpredictable effects on their ecology and welfare.

As an expression of concern about the ethical implications of genetic engineering and its profound effects on humanity, animals and the environment, a majority of those attending the Athene Trust International Conference endorsed a resolution to call for the banning of the transgenic manipulation of animals and the patenting of genetically engineered living organisms (see 'Conference Resolutions').

Regulation

Murphy expresses the consensus of all the contributors to this part of the book when she states that genetic engineering is an extremely powerful technology that must be the subject of humane legislation, open debate and responsible controls. In Bulfield's opinion, the new techniques of genetic engineering are being carefully regulated in comparison with conventional animal breeding techniques. The UK guidelines on work with transgenic animals, issued by the Health and Safety Executive (HSE) in January 1989, make it compulsory for anyone intending to produce transgenic animals, or intentionally introduce them into the environment, to notify the HSE (ACGM/HSE/Note 9; see also Chapter 23). However, the control of product safety and animal and plant health issues are outside the HSE's remit. This is because the HSE derives its authority from the Health and Safety at Work (HSW) Act (1974), which charges an employer to provide a safe working environment for employees and to avoid exposing the general public to risks. The HSW Act does not provide for the enforcement of matters concerned with ecological or animal welfare effects if no adverse effect on human health can be established.

The release of animals of a kind not ordinarily resident in, and not a regular visitor to, Great Britain is legitimately regulated by the Department of the Environment (DoE) under the Wildlife and Countryside Act (1981). However, it is not certain whether transgenic animals would be considered sufficiently non-native to be regulated under the Wildlife and Countryside Act (1981).

From the perspective of animal welfare legislation, the transgenic manipulation of animals is poised between being an experimental procedure and a farming practice. The situation in the UK is that in the case of farming practices the welfare of transgenic vertebrates other than humans is the responsibility of the Ministry of Agriculture, Fisheries and Food (MAFF), and in the case of experimental procedures it is the responsibility of the Home Office under the Animals (Scientific Procedures) Act (1986) (A (SP) Act).

The A (SP) Act regulates any experimental or other scientific procedure applied to a protected animal which may have the effect of causing that animal pain, suffering, distress or lasting harm. The transgenic manipulation of animals is perceived to be on the borderline between being a scientific procedure – and therefore subject to the Act – and a recognised veterinary, agricultural or animal husbandry practice, in which case it would be exempt from the Act. The insertion of DNA into the germline, which is part of transgenic manipulation, and the subsequent breeding of transgenic animals, are presently classified as experimental procedures within the meaning of the Act (see also Chapter 23). Therefore, in the UK at present, all proposed transgenic manipulation work with vertebrates requires a Project Licence and a Personal Licence from the

Home Office. However, were the procedures to come to be regarded as routine procedures, they would no longer be covered by the A (SP) Act. Similarly, once produced, if transgenic animals could be shown to be normal in all respects apart from the genetically engineered characteristic, they would no longer be protected under the Act, and would be subject only to the same welfare controls as other farm and laboratory animals. These controls include the Protection of Animals Act (1911) (in Scotland, 1912), the Agriculture (Miscellaneous Provisions) Act (1968), and the Transit of Animals Orders (1973 and 1975).

In the USA, where animal welfare policies are exercised by the Public Health Service (PHS), the National Institutes of Health (NIH) and the Department of Agriculture's (USDA's) Animal and Plant Health Inspectorate Service (APHIS), there are both gaps and overlaps. Fox points out that the two major groups of animals utilised in genetic engineering studies – livestock and rodents – together with birds, fish, reptiles and amphibians, are excluded from protection under the Animal Welfare Act (AWA) as administered by the Veterinary Services of the APHIS. Moreover, to the dismay of the HSUS, even the controversial new federal animal welfare regulations, proposed by the APHIS in the summer of 1989, exclude birds, mice and rats (McGourty 1989a). Although livestock and rodents are covered by the NIH animal welfare guidelines, these guidelines are only binding on research which is federally funded.

Participation

The sole remit of governmental agencies like the DoE in the UK and the Environmental Protection Agency (EPA) in the USA is to protect the environment. There is no equivalent government agency whose sole remit is the safeguarding of animal welfare. Animal welfare issues are regulated by a number of agencies whose main functions range from research, the environment, farming and food production, to industry, education and defence. The potential conflict of interests in state agencies which are both sponsors of research on animals and regulators of the welfare of its animal subjects is of some concern. For example, the transgenic experiments conducted by the USDA in Maryland which produced the 'Beltsville pigs' receive unanimous condemnation from all contributors to this part of the book. In these experiments, foreign growth hormone genes were inserted into pig embryos to increase their growth rate. As a result of the transgenic manipulation, the 'Beltsville pigs' were grossly deformed – being crippled with arthritis, cross-eyed and having a wrinkly rust-coloured skin.

In the absence of regulatory agencies dedicated to the protection of the welfare of animals, it tends to be the subject of specialist committees, formed on a local or national level, to advise, and sometimes to adjudicate on procedures and practices which affect animal welfare. In France, for example, there is a national commission which advises the government on all questions concerning animal experimentation. The commission comprises eight representatives from the seven ministries concerned with animal experimentation and twelve experts drawn from both the public and the private sectors (see Coles 1989). In Britain, the MAFF and the Home Office each have a specialist committee – the Farm Animal Welfare Committee (FAWC) and the Animal Procedures Committee (APC) respectively – to advise them with regard to their regulatory duties towards the welfare of farm animals and laboratory animals. Webster, who is a member of the FAWC, has been nominated to represent the FAWC at discussions in the APC on matters of shared concern and interest, including transgenic manipulation (APC 1989).

One of the demands that animal welfare groups have campaigned for, with some success, is the inclusion of members of the public and animal welfare campaigners on the committees and commissions which advise, adjudicate, inspect and sanction procedures and practices which affect animal welfare. The new federal laboratory animal welfare regulations, issued by the APHIS in the USA in 1989, stipulate that there must be a member of the public on the institutional committees that will monitor compliance with the regulations (McGourty 1989b). Similarly, under the laboratory animal welfare ordinance introduced by the City of Cambridge, Massachusetts, in 1989 there must be a member of the public on the animal-care committee formed at each research institution. These autonomous committees are empowered to prevent or restrict experiments. However, a proposal that there should be an animal welfare campaigner on each animal-care committee was vociferously resisted by the research community and rejected by the City Council (Shulman 1989a; 1989b).

The objection usually raised against the inclusion of animal welfare campaigners on committees and commissions which advise, adjudicate, inspect and sanction procedures and practices which affect animal welfare is that such individuals cannot be trusted to act in a reasonable and constructive way (see for example, Shulman 1989a; 1989b; Hansard 1985). However, the idea that animal welfare campaigners are to be less trusted to act constructively than those who have a vested interest in the procedures or practices under scrutiny was not shared by the British government when they passed the A (SP) Act in 1986. Section 19 of the Act prescribes the composition of the Animal Procedures Committee (APC) which must be appointed to advise the Secretary of State for the Home Office on matters concerned with the Act. It stipulates that not only must the interests of animal welfare be represented by the membership of the APC, but also that at least half of the members of the APC, other than the Chair, must be persons who neither hold, nor have held within the previous six years, a licence to carry out scientific procedures on animals.

It seems incredible that a quarter of a century ago government departments and agencies dedicated to the protection of the environment did not exist. It will perhaps seem just as incredible to the next generation that today there are no government departments and agencies dedicated to the protection of animal welfare. In the absence of such agencies, the specialist committees charged with protecting the welfare of animals subjected to transgenic manipulation have a very important role to play. The balance of membership between animal researchers, veterinarians and animal welfare campaigners who serve on the committees that regulate the transgenic manipulation of animals is of crucial importance both to safeguard animal welfare and to assure the general public that animal welfare interests are being adequately taken into account. It is essential that scientists sitting on such committees declare any vested interests they may have in this type of work and that at least 50 per cent of the membership should *not* be involved in the transgenic manipulation of animals.

The general consensus of the contributors is that openness and public participation can be constructive elements in the regulation of animal procedures and practices for safeguarding the welfare of animals and for achieving greater public confidence in the integrity of the scientific endeavour.

2 Genetically Engineered Animals

Dr CAROLINE MURPHY

Mythical Creatures

Making animals that are a mixture of different species is not a new idea, but it is a new reality. One consequence of this is that there is a common ground that we all share when we start thinking about the type of genetic engineering that we call the transgenic manipulation of animals. Here I describe briefly what transgenic manipulation is, and point out some implications which may be both reassuring and disconcerting.

For many people, the transgenic manipulation of animals is a very frightening concept. The reason why we should find this application of modern genetics so disturbing lies at least partly in our cultural heritage. The fantasy of animals with a mix of characteristics of various species can be found in the culture of many ancient, and not so ancient, civilisations. When trying to decide what our reactions to transgenic animals should be, we must be aware that many of us will have seen, or chosen not to see, feature films about mad scientists who have shut themselves away from the world and set about producing half-human, half-beast monsters. Mary Shelley's *Frankenstein* nightmare has produced many abiding cultural images and exposed deep-seated fears about the consequences of mankind's interference with nature. When we look at transgenic animals we must be aware that our reactions to them are unlikely to be based on the rational analysis of facts ably presented; the nightmare images from past fantasies are too likely to escape from the Pandora's Box of our imagination and distort our vision.

Transgenic Manipulation

Transgenic manipulation is a type of genetic engineering. It can loosely be described as taking genes from one organism and inserting them into the genetic material of another. A gene is a length, or a series of lengths, of DNA which carries the code for the sequence of amino acids that make up a protein. Every different protein has a unique amino acid sequence: that sequence is coded for in a gene. The genes lie interspersed with regulatory DNA in a fixed order along the chromosomes. There are gene-bearing chromosomes in the nuclei of all actively dividing cells and they are large enough to be seen under a light-microscope in a dividing cell nucleus. A gene is too small to see with the light-microscope, but banding patterns, generated by staining techniques, help us to 'map' the location of genes on chromosomes fairly accurately. For example, using a variety of gene mapping techniques, we have identified that the gene for the blood protein alpha globin is, in humans, on chromosome 16. In transgenic manipulation, a gene from one species relocates permanently – the scientists hope – into one of the chromosomes of another species.

When a sperm fertilises an egg, the chromosomes of the mother and father pair up to produce the unique genetic characteristics of the offspring. Today's genetic engineering of transgenic animals involves the insertion of relatively small amounts of DNA. The amount of foreign DNA inserted into the chromosomes is so small that it does not interfere with this chromosome pairing process. Therefore transgenic animals are able to interbreed with each other and with other (non-transgenic) members of their species. It is hoped that the foreign gene, like any other gene, is passed on to about half the offsring. The normal transmission of half of each parent's genes to each offspring can apply to transgenic animals so successfully that, by normal mating of two different transgenics, young are born carrying both transgenes (Goodnow *et al.* 1988).

The first experiments with transgenic manipulation were undertaken with bacteria and viruses. The commercial application of this work offers possibilities to people and animals. Bacteria carrying the gene for making the human insulin protein have already been produced. These bacteria have been cultured to produce commercial quantities of human insulin for use in place of insulin derived from pig or cow pancreases. However, many patients find it harder to control their diabetes using the new genetically engineered human insulin than using the insulins extracted from pigs or cows.

Nowadays transgenic plants and animals are being created in the laboratory. If wheat could be genetically engineered to fix nitrogen from the air (as the legumes can, naturally), wildlife could benefit from reduced run-off from nitrogenous fertilisers into streams and rivers.

But what about the genetic engineering of animals? The science magazine *Nature* introduced the reality of transgenic animals to the world with a spectacular front cover (*Nature* 1982) [see also Chapter 3]. The cover depicted two apparently perfectly healthy mice; one of normal size, together with a 'giant' transgenic sibling with the growth hormone gene of a rat in every cell of its body. This 'giant' mouse was one of only seven transgenics produced from 170 specially treated egg cells. The researchers who had genetically engineered these transgenic mice nonetheless recognised that: 'the implicit possiblity is to use this technology to stimulate the rapid growth of commercially valuable animals. Benefit would presumably come from a shorter production time and possibly from increased efficiency of food utilisation' (Palmiter *et al.* 1982).

The pigs carrying human growth hormone genes produced by USDA scientists at Beltsville in Maryland, USA, were among the first and most widely publicised transgenic animals. These pigs expressed high levels of growth hormone and developed severe arthritic deformities, probably as result of the regulatory gene sequence attached to the human growth hormone gene (Hammer *et al.* 1985; Pursel *et al.* 1987) [see also Chapter 3].

Applications

There are three main types of genetic engineering of vertebrates being carried out at the moment. These are the production of animal disease models for use in medical research; the production of pharmaceutically useful human proteins in animals, and the production of 'improved' agricultural animals. Each of these applications of genetic engineering may have both a positive and a negative side. The sheep which produce Factor IX in their milk are fit and healthy (Hammer *et al.* 1985; Pursel *et al.* 1987) [see also Chapter 3]. Factor IX is the human blood-clotting factor missing in a type of haemophilia called 'Christmas disease'. If sheep could similarly be genetically engineered so that they lactate Factor VIII, the absence of which is the most common cause of haemophilia, haemophiliacs the world over could benefit.

Much valuable grazing land in Africa cannot be used by farmers because of tsetse flies which carry various trypanosome parasites that cause diseases in cattle and humans, notably 'sleeping-sickness'. If cows could be genetically engineered to be resistant to these parasites, the farmers would be able to graze more land. In the short term, this might be good for the farmers and their stock, but the wild animals which are already resistant to 'sleeping-sickness' would be deprived of undisturbed grazing land, and their numbers would be bound to fall further as they were pushed to more marginal land.

The first patented mouse had a human cancer gene inserted into its genome so that it could act as a more 'accurate' model of human cancer than the mice that were already used (Ezzell 1988;

Joyce 1988). These patented mice are already on the market and were first imported into the UK in the spring of 1989 (Ezzell 1989).

In my opinion one of the most distressing applications of transgenic manipulation is the production of mice that have uncontrolled growth of the crystallin protein in the lens of the eye. Prior to the production of this disorder using genetic engineering no cancer of the lens of the eye of any animal had ever been observed. Now there is a strain of laboratory mice that have to have their eyes removed if they are to be used for breeding. If their eyes are left in they inevitably develop a fatal invasive cancer (Mahon *et al.* 1987).

Poultry producers, worried about the spread of infectious disease through the large flocks of intensively reared birds, are very interested in the prospect of transgenic birds resistant to highly infectious viral infections, such as 'Newcastle Disease' (avian influenza). Those concerned about the birds' welfare question the humaneness of battery production systems. The 'Trojan horse' of disease resistance may provide a means whereby genetic engineers can design animals to cope with conditions that no animal, genetically engineered or not, should be expected to endure.

Another proposal is the production of cows which do not metabolise and break down their naturally produced growth hormone protein – bovine somatotropin (BST). Such cows could be genetically engineered so that they contain extra copies of the BST gene. The consequence of this would be cows, possibly patentable, producing high milk yields due to abnormal levels of their own BST.

There is another particularly worrying possible application of genetic engineering to animals: the improvement of sporting animals and pets. The prospect of patented Derby runners, Olympic show-jumpers, Crufts champions and new breeds for the cat fancier is not one I look forward to. In dog-breeding traditional methods of intensive selection and repeated in-breeding have produced animals that have chronic eye and breathing problems, reduced life expectancy, and need routine surgical assistance at whelping. In the nineteenth century such genetic 'oddities' were called 'sports'. In my opinion the creation of 'sports' is not a suitable application of genetic engineering.

In this chapter, I have drawn attention to the important cultural role of mythical beasts – human–animal and animal–animal hybrids – in providing us all with nightmarish preconceptions about the crossing of species barriers. Scientists and welfarists alike must guard the Pandora's Box of our imaginations in assessing future realities. The Pandora's Box of the human imagination tempts scientists to out-do Doctor Frankenstein, and tempts welfarists to shut the lid on what may offer a cornucopia of opportunities. We must neither assume that all genetic engineering is evil, nor be seduced into thinking that the bright light of science can offer only progress. The genetic engineering of animals is here to stay. It offers both possibilities and problems that we must address intelligently. While the mythical 'transgenic' creatures of the past were thought to control our destiny, the power to control the genetic engineering of animals lies in our hands. We must ensure that we regulate and control this new technique to the benefit of both ourselves and other animals.

3 Genetic Manipulation of Farm and Laboratory Animals

Dr GRAHAME BULFIELD

On the front page of the journal *Nature* (December 1982) was a picture of two littermate mice, one twice the size of the other. The larger animal, when it was a single-celled embryo, had been injected with about 600 copies of an exogenous growth hormone gene (Palmiter *et al*. 1982). The gene was present in every cell in the animal, was passed on from generation to generation and produced about 1,000 times more growth hormone than the mouse's normal gene. The mouse grew to twice the normal size. This experiment stimulated many scientists working in agriculture and animal breeding to become interested in the techniques of genetic manipulation because, for the first time, an agriculturally important characteristic – growth – had been manipulated.

In this chapter I shall describe in simple terms the techniques used in producing these genetically manipulated, or 'transgenic', animals and explain how they are likely to be useful, and finally I shall discuss the animal welfare and other implications of this technology.

The procedures for producing transgenic mice are well established (Palmiter and Brinster 1986). The single-celled fertilised mouse embryo is flushed from its mother's oviduct. The DNA of a single growth hormone gene is isolated and cloned in bacteria so that large quantities of it are produced. It is then injected into a pronucleus of the mouse embryo under a microscope. The embryo is then implanted into a surrogate mother which will give birth to offspring, some of which will be transgenic, that is they will contain the exogenous growth hormone gene in all of their cells. The proportion of embryos injected that are transgenic is about 3 per cent but, as most of the loss is before birth, the proportion of transgenic animals amongst live-born offspring is much higher, between 15 and 25 per cent (see also Chapter 5). These techniques have been successfully applied to produce transgenic pigs, sheep and cattle (Brem *et al*. 1986; Simons *et al*. 1987).

There are three areas in which transgenic animals have been

used. The first is in investigations of gene expression in the control of physiological systems and the regulation of animal development. The second area is the breeding of commercially superior animals and the third area is the secretion of pharmaceutically important compounds for human use in the milk or other tissues of transgenic animals.

For the first time we can put a single exogenous gene into a mammal, control its regulation and alter its effect in different tissues. Therefore, genetic manipulation is an important technique for investigating the biology of higher organisms and will revolutionise mammalian biology. This is exemplified in the large transgenic mice. The gene injected into these mouse embryos was actually a hybrid between a growth hormone gene and a regulatory sequence from another gene that functioned in the liver. Once inside the transgenic mouse, this hybrid gene did not produce growth hormone in the pituitary gland – the normal tissue that produces growth hormone – but produced it in the liver. The growth hormone was produced in totally foreign tissue, exported properly into the blood, and it was biologically active.

These transgenic mice demonstrate that regulatory sequences, or 'switches', from one gene which are attached to another gene will still work correctly – something that was unknown prior to these experiments. The production of growth hormone in the liver of the transgenic mice was not part of the feedback loops that regulate normal growth hormone synthesis. The hybrid gene was switched on by giving the animals zinc chloride in their drinking water because the regulatory switch attached to the growth hormone was also sensitive to heavy metals. So the inserted gene was not only switched on in an incorrect tissue, but could be regulated by external factors, which means the gene is under our control.

Farm Animals

There are two problems which prevent the routine application of transgenic manipulation techniques to farm animals. The first is the identification of the genes involved in commercially important traits. The second is inserting the genes and regulating their expression in the right tissues at the right time. The latter problem has now been overcome with all farm animals except poultry. The identification of trait genes is, however, still a serious problem. There are about 100,000 expressed genes in an animal, and most commercially important traits, like growth, fertility and milk production, are controlled by very many genes. We do not know the number or nature of these genes, their chromosomal location or their products. In fact we only know the functions of about 10 per cent of all genes. Therefore, before genetic manipulation can be used widely in farm animals a lot more research is needed to try to understand the physiology of traits like muscle growth or lactation. This is a serious problem which will inhibit the widespread use of genetic manipulation in farm animals.

The genetic manipulation techniques developed for mice have been adapted for sheep, pigs and cattle. However, in these farm animals these techniques are still very expensive and logistically complicated to execute. The success rate for the transgenic manipulation of farm animals is similar to mice (1–3 per cent of embryos injected).

The transgenic sheep produced by the AFRC Institute of Animal Physiology and Genetics Research Station in Edinburgh provide an example of the use of animals as bioreactors producing pharmaceutically important compounds in their milk. This has been achieved by linking part of the gene for the sheep milk protein lactoglobulin to the human blood-clotting Factor IX and then using this hybrid gene to create transgenic sheep. The Factor IX protein can then be recovered from the sheep's milk (Clark *et al.* 1989). Patients with a particular type of haemophilia are deficient in Factor IX. They are usually treated with Factor IX extracted from human blood, which unfortunately can be contaminated with pathogenic (disease-causing) viruses. Production of Factor IX from transgenic sheep overcomes this problem. Whether this method of producing human proteins becomes routine or not will depend on the availability of other sources. With regard to animal welfare, the transgenic sheep produced so far are normal; the exogenous gene does not appear to have any deleterious effect on the animals (Simons *et al.* 1987).

Existing transgenic manipulation techniques have not proved suitable for poultry. The main problem is that the laid egg is a 24-hour embryo of about 60,000 cells. One way of producing transgenic poultry is to use a total *in vitro* culture system for chicken embryos from fertilisation to 'hatch' (Perry 1988). Using this technique, which has been patented, both the cultured embryo and the 'hatched' chickens develop normally and the embryo is available for manipulation throughout development. So far, however, a transgenic chicken has not been produced.

In summary, what has been achieved is to establish routine genetic engineering techniques which work as successfully on farm animals (except poultry) as they do on laboratory animals. The success rate is still not very high and the procedures are logistically complicated and expensive. Moreover, transgenic manipulation cannot be used in animal breeding practice until we identify and understand the nature of the genes controlling commercially important traits.

Animal Welfare

I now wish to place the techniques used for creating transgenic animals into a wider context. Genetic manipulation techniques put one, or at the most two, extra genes into an embryo. There are about 100,000 expressed genes in a mammal, so we are changing the genetic make-up of the animal very slightly – about 0.001 per cent. Human beings have already changed the genetic make-up of all domesticated animals dramatically. The modern Holstein Friesian cow looks nothing like the European bison; a modern Landrace pig looks nothing like a wild boar, and there are tremendous variations between breeds of dogs. In these cases breeders have changed the genetic make-up of animals by hundreds or even thousands of genes and it has been done in an uncontrolled 'black box' manner. As a result of selective breeding deleterious secondary effects have often occurred with farm animals as well as with dogs and cats.

Genetic engineering places a powerful but very controlled tool in the hands of breeders. Using genetic engineering we can alter one gene at a time and keep everything else constant. If the gene turns out to have deleterious effects it can be rejected and not used in further experiments. Therefore, instead of negatively affecting the welfare of animals, genetic manipulation should bring breeding under the control of the scientists and have a positive effect on animal welfare.

There has been a lot of discussion about the arthritic transgenic pigs produced by the USDA that contained a human growth hormone gene. I would like to comment on the use of the human growth hormone gene first, and then on the arthritis produced.

Human growth hormone contains about 250 amino acids, and it differs from pig growth hormone or cattle growth hormone by just a small number of amino acids. At the molecular level there is very little difference between these growth hormones; we could even 'create' a human growth hormone gene from a pig growth hormone gene by altering the pig gene in a test-tube. Unless the human gene used in the USDA pig experiment was very similar at a molecular level to the pig growth hormone gene it would not have functioned in the pig. I do not know why a human gene was used in this experiment since I would have expected a pig gene to work more efficiently.

A breed of pigs could exist that are arthritic as a result of a naturally occurring mutation, pigs could be made arthritic by treating them with drugs, or we could make them arthritic through transgenic manipulation. Whatever the cause of the arthritis, whether it be drugs, a naturally occurring mutation, genetic manipulation or some other cause, if pigs are arthritic and they are suffering it should not be permitted.

Work on transgenic animals in the UK is controlled by a number of acts and regulations. The Advisory Committee on Genetic Manipulation (ACGM) of the Health and Safety Executive (HSE) published guidelines on transgenic animals in January 1989 [see Chapter 23] and proposals for genetic manipulation projects in the UK must be submitted to the ACGM. In addition to the ACGM guidelines, the Animals (Scientific Procedures) Act (1986), administered by the Home Office, covers work with transgenic animals. Any transgenic animal with clinical side effects that is produced would be regulated under the Act and would not be allowed to be released from the laboratory for animal breeding purposes. The Ministry of Agriculture, Fisheries and Food (MAFF) and the Department of the Environment (DoE) also have regulations governing farm animals and their welfare.

The ACGM Transgenic Animals Working Party was established, not because of the RSPCA, and not because of the animal welfare lobby, but at the request of scientists working in the area. The scientists approached the ACGM and requested guidelines to regulate their work, and the Transgenic Animals Working Party was established. This working party had representatives from the Home Office, the DoE and the MAFF, as well as scientists.

I should like to give two examples which illustrate the tight controls and regulations governing work with transgenic animals in the UK. A breed of cattle called the Belgian Blue has a naturally occurring mutation that causes double muscling of the hind limbs. If Belgian Blues are crossed with certain breeds of cattle serious calving problems occur. However, there is no legislation to prevent such cross-breeding. If we had produced the Belgian Blue mutation by genetic engineering, Belgian Blues would not be permitted to be used in the field and would have to remain under the Animals (Scientific Procedures) Act (1986). Similarly, if an arthritic pig were to arise naturally (that is by naturally occurring mutation), it could be used commercially. If it were to be genetically engineered in the laboratory, it could not be released.

Genetic Manipulation of Humans

Another area that causes concern is the possibility of the genetic manipulation of humans. It is highly unlikely that for medical or other reasons scientists would want to manipulate the human germline. If you select mice for 25 generations for weight they end up being three times larger than unselected mice. People have been breeding domestic animals for 8,000 years. Genetic manipulation through selective breeding has therefore been available for a very long time. However, no society has ever permitted selective breeding practices on human beings for any length of time.

It may be desirable to replace a mutant gene in the tissue of a patient with an inherited genetic disease. For example genetic manipulation might be used to target a normal gene to the muscle in a child suffering from Duchenne muscular dystrophy. Duchenne muscular dystrophy is a form of progressive wastage of the muscles which affects one in 3,000 boys. The mutant gene in Duchenne muscular dystrophy has now been isolated and cloned. If we could find some way of targeting a normal version of that gene to the muscle of children with the disease we might be able to alleviate its symptoms.

Genetic manipulation of this kind which involves the insertion for therapeutic purposes of exogenous genes into cells other than the sperm, the egg or the embryo, is known as 'somatic cell gene therapy'. Whilst we are quite a long way from achieving this kind of therapy, it would be similar to organ transplantation. A kidney transplant is, in effect, the transplantation of millions of genes from one person to another. In somatic cell gene therapy, only one gene would be transferred and targeted to operate in a specific tissue – the tissue where it was required. The germline of the patient would not be affected and the manipulation would not alter the genetic make-up of the patient's children. I consider that this type of genetic manipulation of humans can only be beneficial and should present no unique ethical or other objections.

Conclusions

In this chapter I have reviewed the techniques for the genetic manipulation of laboratory and farm animals and have discussed their applications. I have pointed out that there have been some problems in adapting transgenic techniques to farm animals and that these problems have yet to be overcome with poultry. I have described the UK regulations which address genetic engineering and animal welfare and have considered the scope of protection for genetically engineered animals compared to animals which have not been genetically engineered. My conclusion is that, compared with conventional animal breeding techniques, the new techniques of genetic manipulation are precise and controlled and are being carefully regulated.

4 Animal Welfare and Genetic Engineering

Professor JOHN WEBSTER MRCVS

The bio-revolution raises fundamental questions of ethics and theology and practical problems of consumer acceptability. As a contribution to the ethical debate I should like to quote the following passage from the Reverend Carpenter's Working Party on Animals and Ethics: 'The degree of suffering experienced by an animal is dependent upon its own physiological and anatomical make-up and is totally unrelated to its beauty, its rarity, its economics or its nuisance value' (Carpenter 1980). Animal welfare, as perceived by the animal, implies, at the very least, an absence of suffering. This is a property of the animal itself and applies whether we wish to eat it, pet it, or experiment with it – either justifiably, as in the case of cancer research, or unjustifiably, as in the case of cosmetics testing. It is all the same to the animal.

The Farm Animal Welfare Council (FAWC), in their *Report on Priorities in Animal Welfare Research and Development* (FAWC 1988), stated: 'We accept that scientific investigation aims to be impartial and without prejudice so that it is impossible to pronounce *a priori* how any particular piece of new knowledge will affect the welfare of farm animals'. Nevertheless there are certain areas of research and development which, if applied in practice, could possibly lead to a significant deterioration in animal welfare. These include the manipulation of body size, shape or reproductive capacity by breeding, nutrition, hormone therapy or gene insertion in such a way as to reduce mobility, or increase the risk of injury, metabolic disease, skeletal or obstetric problems and perinatal mortality.

The application of genetic engineering to animal production may be considered under three headings:

Direct Engineering of the Genotype This includes manipulation of body size, shape, reproductive performance and the composition of meat and milk. It also includes genetically engineering animals for improved disease resistance.

Use of Genetically Engineered Products This includes genetically engineering plants to improve the supply and quality of animal feeds; engineering microbes to improve the availability of feeds; growth and lactation promoters like recombinant (genetically engineered) growth hormone; vaccines, and diagnostic aids.

Acceleration of Genetic Change This, which is really an adjunct to genetic engineering, includes multiple ovulation with embryo transfer (MOET); cloning of 'improved' livestock, and the production of chimeras (new types of hybrid animals).

For many scientists these applications of genetic engineering are very exciting, but what can they hope to achieve?

There are, I think, only five ways in which one can improve animal production, and these are:

(i)   Improve the supply or quality of animal feed or reduce its cost.
(ii) Improve the conversion of animal feed into saleable product, for example meat and milk, in the animal that eats that feed.
(iii) Improve reproductive efficiency so as to reduce the overhead cost of maintaining the parent population.
(iv) Reduce losses through improved animal health.
(v)   Manipulate the composition of animal produce.

I shall consider a few examples to illustrate the potential of genetic engineering to manipulate these factors and the implications thereof for farm animal welfare.

Manipulation of Feed

There are very sound biological reasons to justify the claim that, in ruminant species at least, the greatest gains have been achieved through improvements in nutrition. Most nutritional work may be considered as 'welfare-neutral'. There are several areas where nutrition may appear to impinge on welfare problems, for example metabolic disease in dairy cattle consuming large quantities of concentrate feed. However, on closer inspection, food is not the primary culprit. Appetite in farm animals is driven largely by the need to meet the physiological capacity of that animal to grow or lactate. When we manipulate that capacity, either genetically or by use of exogenous hormones, then the animal eats to keep up.

Manipulation of the Animal Genotype

Some of the most dramatic claims and counter-claims relating to genetic engineering concern transgenic animals containing a foreign growth hormone. Thus manipulated, mice and pigs grow twice as big as normal, but in both species there have been severe skeletal and arthritic problems which make the practical application of the project quite unacceptable on welfare grounds. The people who undertook this work cannot be blamed for failing to anticipate the arthritis, but if they had had a sounder knowledge of animal science they would have known that there is no direct relationship between size and efficiency. For example, Falconer (1973) at Edinburgh selected mice for and against growth rate so that after 15 generations one strain was twice as big as the other. This, however, had absolutely no effect on the *efficiency* of growth.

In fact it is not even necessary to quote experiments of this sort to demonstrate that bigger is not necessarily better. If it were, the cow would be 200 times as efficient as the chicken. In practice it appears that through genetic selection, the broiler chicken has become much more efficient than the lamb or beef calf. The spectacular improvements in growth rate, carcass composition and efficiency of food conversion in pigs and poultry have been achieved largely by selecting for animals that are bigger when adult, and killing them at an increasingly early stage. Extrapolating this argument towards absurdity, the infinitely efficient animal is one killed at birth. The obvious flaw here is that income from meat would be tiny compared with the cost of maintaining the parent population. When the offspring slaughtered for meat eat most of the food, as with poultry and pigs, it does pay to select for bigger and yet bigger animals, and to kill them when they are relatively small compared to their mothers. When the proportion of food consumed by parent and offspring is 50:50, as with cattle, the benefit of size disappears.

Manipulation of size by genetic engineering appears neither very sensible nor very new as anyone might reasonably deduce from a visit to Crufts Annual Dog Show. However, even conventional genetic selection for traits such as growth rate and carcass composition to the exclusion of other normal body functions, such as mobility, can create welfare problems. Crippling deformities of the limb bones and joints are becoming acceptable as normal in broiler chickens. If animal welfare matters, then the conventional genetic selection of broiler chickens for rapid growth has already gone too far. Broiler lameness could be overcome either by selecting a less fast-growing strain or by restricting food intake. Animal health overall would be improved, and so would the performance of some individuals, but the producer would lose time and money. We have here one of many exceptions to the common argument that productivity and health are synonymous, or 'if they weren't happy they wouldn't grow so well'. This exception creates a general rule:

When assessing any new development in animal production, whether achieved by genetic engineering or not, one cannot accept optimal productivity of the unit as sufficient evidence for adequate welfare of the individual.

The manipulation of animal shape and composition is, in biological terms, a more realistic approach to improved productivity than manipulation of size. I can see no objection in principle to the manipulation of shape unless it creates pain or discomfort, impairs mobility or interferes with other essential body functions. By this reasoning it is acceptable to enlarge the breast muscle of the turkey to the point where normal mating becomes impossible, but unacceptable to manipulate the muscles and pelvic dimensions of the Belgian Blue cow to the point where she can no longer give birth by natural means.

Genetically Engineered Hormones

Most current concern is centred on the use of genetically engineered porcine or bovine growth hormone (PST or BST) [see Part II] to manipulate growth in pigs, or lactation in dairy cows. We must ask: is this a 'procedure which may cause pain, suffering, distress or other lasting harm due to disease, injury, physiological or psychological stress, significant discomfort or any disturbance in normal health, whether immediate or in the longer term'? These are the words used to describe a regulated procedure within the Animals (Scientific Procedures) Act (1986) designed to improve the welfare of experimental animals. Before Home Office approval under the act, the predicted severity of any procedure must be balanced against the potential benefit of the work. If the regular administration of growth hormone 'in amounts only slightly outside normal physiological limits' is shown to have no significant deleterious effects, then it would be defined by the 1986 Act as a mild procedure. If there are other, as yet unforeseen, effects then the severity will become more substantial. By analogy, when considering the application of growth hormones in agriculture, we must include some form of cost-benefit analysis. In the case of BST we have a product which the consumer in developed societies does not want, the poor in developing countries cannot use, a product which no dairy farmer actually needs and which will drive some out of business. Given these 'benefits', the costs must be assessed very severely indeed.

My main objection to the use of BST is that the dairy cow, like the broiler chicken, is already at her metabolic limit. I have explained elsewhere how the work load of the dairy cow exceeds that of any other farm animal and greatly exceeds that of humans engaged in hard manual labour, such as miners and foresters (Webster 1988). Very recently, however, Dr Westerterp in the Netherlands has acquainted me with a group of humans who do work as hard as the dairy cow. Cyclists engaged in that supreme exercise in masochism, the Tour de France, work at a rate equivalent to that of a Friesian cow yielding about 35 litres of milk per day. I would add that drug-taking by humans is frowned upon in these circumstances!

Manipulation of Reproductive Efficiency

The most promising application of biotechnology to the manipulation of reproductive efficiency is multiple ovulation with embryo transfer (MOET). MOET is a technique whereby multiple eggs are removed from a beef cow after slaughter, fertilised in a test-tube with semen from a beef bull, sexed, and the fertilised embryos implanted into a recipient cow. The commercial advantages of this are clear. A dairy cow may be implanted with one or two of the beef-type calves. Similarly she may be implanted with female dairy-type calves of superior genetic merit to herself. The procedure is non-surgical and, so far as the recipient cow is concerned, involves no greater indignities than artificial insemination. I can see absolutely no objection to this procedure on welfare grounds. Indeed, if we are to rear very heavily muscled Belgian Blue type calves (and the biological justification for such animals is extremely dubious) then it will be more humane for them to undergo gestation in a normally framed Friesian cow which can deliver them without assistance than in a Belgian Blue which will require caesarean section.

Animal Health

At first sight the potential for genetic engineering to improve health appears almost limitless through improved vaccines, diagnostic aids and the insertion of genes to confer disease resistance. In fact most of the diseases of farm animals in the developed world that are amenable to these approaches are already reasonably under control. The diseases that remain are either non-infectious consequences of existing husbandry systems, for example metabolic disease and lameness in dairy cows and musculo-skeletal problems in chickens, or infectious diseases such as pneumonia and mastitis which are multifactorial in origin and where the vaccinal approach has been singularly unsuccessful.

I forecast, therefore, that genetic engineering is likely to have little positive effect on farm animal health, partly because it is not appropriate to most economically important production diseases and partly because a genetically engineered product that produced an instant cure in a few farm animals would not repay the development costs. Thus the National Office of Animal Health (NOAH), a most Orwellian title, promotes BST, in my opinion, not in the interests of animal health but because pharmaceutical companies hope to be able to inject it into nearly all the cows nearly all the time and so satisfy their shareholders (see also Chapter 8).

Manipulation of the Composition of Animal Produce

This, to my mind, is potentially the most exciting and most humane application of genetic engineering to animal production. Ideas under investigation include the incorporation of human genes, such as the gene for blood-clotting Factor IX, into sheep in such a way that the protein for which the gene is a blueprint can be secreted in sheep's milk [see Chapter 3]. The human protein could then be isolated from the sheep's milk and used: for example, Factor IX protein can be used for the treatment of certain types of haemophilia. Even more recently, genetically engineered strains of organisms which cause food poisoning in humans have been used to vaccinate cows which then secrete antibodies in their milk which may be used for the prevention and treatment of food poisoning. It is difficult to see how the application of these techniques could cause animal suffering, quite the reverse, since each successfully manipulated animal would become very valuable indeed, and I say without cynicism that is the best way to ensure its welfare.

To make a more general point, it is neither imaginative nor in the interests of farm animals to adopt the entrenched position that they only serve to provide food, work and clothing. The sheep or cow that produced pharmaceuticals designed to reduce human suffering would not, I think, be abused or neglected.

Conclusions

To return to the fundamental question, 'Will genetic engineering provide a cornucopia or open a Pandora's Box?' The simple answer is 'neither'. For a start there is a limit to what can be achieved by insertion or deletion of a single, or small number of genes; secondly, genetic change is not synonymous with genetic improvement and, finally, animals are much more similar than they appear. Thus the scope for useful change is less than its advocates or detractors might think.

I do not believe that the genetic engineering of farm animals is so different from other advances in nutrition, breeding and management that it poses unique questions relating to animal welfare. In this chapter I have considered a range of applications of genetic engineering to animal production and come to some widely differing conclusions. This implies that each case should be treated on its merits. I would, however, make one final general point: genetic engineering is not a new science, it is only a novel technique and we must not let ourselves become seduced by the excitement of novelty. Whenever science provokes the response 'Gee whiz!', we must immediately add the caveat 'So what?' The 'geep' – a half-sheep, half-goat chimera – is a classic example of 'Gee whiz; so what?' When we do ask 'So what?', it must be in terms both of sense and sensibility so that we may learn to distinguish between the realistic, the unlikely and the unfair.

5 Transgenic Animals: Ethical and Animal Welfare Concerns*

Dr MICHAEL FOX

Transgenic manipulations can be used by genetic engineers to make farm animals more productive, to turn animals into 'pharmaceutical factories', and to develop laboratory animal models for various human diseases. However, such manipulations raise a host of ethical and animal welfare concerns. This chapter considers the ethics and consequences of using genetic engineering to create transgenic animals.

The *telos* [Greek 'end'] of an animal is its nature or 'beingness'. In other words, the 'birdness' and unique qualities of a canary or eagle, the 'wolfness' of a wolf and the 'pigness' of a pig. Transgenic manipulations alter the *telos* of an animal. From a utilitarian perspective and on the basis of the historical precedent of animal domestication, creating transgenic animals is ethically and morally acceptable. But from the reasonable and reasoned perspectives of respect and reverence for life, for the integrity of species and for the sanctity of the individual, the creation of transgenic animals for human profit is morally and ethically unacceptable.

What constraints are there, if any, to limit transgenic changes in other animals? Certainly in the USA there are no legal constraints or even guidelines, although some may feel that we do have a moral obligation to avoid causing animals unnecessary and unjustifiable suffering. But suppose there is no suffering because the animal is made into a barely sentient vegetable, and is thus better able to adapt to life on a 'factory farm'. Is this a humane solution? Some would say yes, that the animal's *telos*, its intrinsic nature or 'beingness', is not inviolable. What is of concern is the actuality of its suffering, and if suffering can be reduced by altering its *telos*, then all well and good. This kind of animal eugenics is consistent with the misapplication of genetic engineering in medicine and

*A version of this chapter was presented at the American Veterinary Medical Association symposium entitled 'Veterinary Perspectives on Genetically Engineered Animals', Washington DC, 19–20 September 1988.

agriculture since it is based upon a false dualistic perception that separates the organism from its environment.

Transgenic manipulations have the potential to harm the animals subjected to them, and the suffering and disease that have already resulted from such manipulations are documented elsewhere in this book. The adverse consequences of transgenic interventions to animals' health and overall wellbeing, and alternatives that make such interventions unwarranted and unjustifiable, are discussed below.

Ethical Limits

Historical precedents, cultural traditions and economic incentives aside, what are the moral and ethical limits to our changing the *telos* of animals? Some consider that the *telos* should be inviolable, since they believe that it is a violation of the sanctity of being to cross the ethical boundary in order to alter the animal for purely human gain. Others consider that we may alter the *telos* of an animal provided that there is no suffering and possibly some benefit to the animal. So they say yes, for example, to keeping birds in cages, to the declawing of cats, and to the ear-cropping of dogs.

The ethical boundary for changing the *telos* of animals is vague and is determined by subjective values and by the ultimate human benefits that may be gained. The benefits or risks to the animals are generally of secondary consideration because of our long cultural history of animal exploitation and our persisting attitude toward non-human animals who are seen as inferior and thus exploitable because human life and needs are considered to be more important than theirs.

The *telos* or 'beingness' of an animal is its intrinsic nature coupled with the environment in which it is able to develop and experience life. We can harm the *telos* in many ways, for example through environmental, genetic, surgical and pharmacological manipulation. To contend that we can enhance the natural *telos* of an animal – and thus by extension believe that we can improve upon nature – is *hubris*. Genetic engineering makes it possible to breach the genetic boundaries that normally separate the genetic material of totally unrelated species. This means that the *telos*, or inherent nature, of animals can be so drastically modified (for example by inserting elephant growth hormone genes into cattle) as to radically change the entire direction of evolution, and primarily toward human ends at that. Is that aspect of the animal's *telos* we refer to as the genome and the gene pool of each species not to be respected and not worthy of moral consideration?

Some scientists contend that animals have no *telos*. Professor M.J. Osborn, Head of the Department of Microbiology in the School of Medicine at the University of Connecticut, has written: 'The idea that a species has a "telos" is contrary to any evidence provided by biology and belongs rather in the realm of mysticism. That mysticism is a poor basis for sound public policy is amply confirmed by history' (Osborn 1984). Similarly, Dr Maxine Singer of the National Institutes of Health (NIH) in the USA opines that:

> History, from Galileo to Lysenko, teaches us that mysticism can never yield rational and wise public policy in scientific matters ... The notion that a species has a telos (a purpose) contravenes everything we know about biology. Species can have, and many in the past have had, a telos (an end), namely, extinction. This is the only telos known to exist (Singer 1984).

The future world that such scientists will create, if they are not restrained, will be one where animals, nature and the entire creative process are controlled and manipulated to satisfy primarily, if not exclusively, human ends.

To change that which is natural is to alter the harmony within living beings and the harmony in their relationship with the external environment. This is the meaning of harm: to cause injury by disrupting the harmony of life. Some altruists reason that it is ethically justifiable to alter the intrinsic nature of an animal, by using genetic engineering for example, to help it to adapt better to the environmental conditions that we impose upon it, such as overcrowding, exposure to disease on the 'factory farm' and poisoning from pesticides and industrial pollutants. They would employ this same technology to help humans adapt better to the harmful environmental conditions that we have created for ourselves and for future generations. But reason informs us that, since the best medicine is prevention, we should correct the causal conditions and environmental factors that lead to disharmony, suffering and disease. The only corrective is not through more laws, regulations, scientific analysis and technical proficiency, rather it lies in living by the principle of *ahimsa*, of avoiding harm to others. Every good doctor knows this as the medical maxim: 'Physician do no harm.' It is surely bad medicine not to address causal conditions and environmental factors for reasons of economics, politics, ideological myopia and historical amnesia. Thus to conclude that there is no ethical objection to changing an animal's *telos per se*, provided the changes do not inappropriately harm the animal, is a contradiction in terms.

Bernard Rollin, the philosopher, writes:

> Given the economic basis of animal engineering, it is plausible that one whose concern is the happiness of animals should show healthy scepticism about *telos* manipulation, for *in fact* it is probably likely to generate more harm than good for the animals (Rollin 1986).

Yet in the same article he endorses the use of genetic engineering in order to change the *telos* of animals in order to make them 'happy in the environments in which we put them'. This endorsement is not simply the altruism of a potentially misleading eugenic idealism – it reflects a false dualism between the organism and its environment. The organism and its environment are one, and we recognise that unity and harmony as health and the full expression of the animal's *telos*. The *telos* is in part preconditioned (if not predestined) for, and dependent upon, a particular environmental niche and optimal conditions for its normal development and expression, which in turn means health and fulfilment for the animal. To deny such health and fulfilment by keeping the animal under impoverished and even stressful environmental conditions (as on a 'factory farm') is to cause harm.

Historical precedence, cultural traditions and economic incentives aside, to engineer animals, as Rollin (1986) suggests, so that they will be 'happy in the environments in which we put them', is totally to disregard the *telos* of the animal: its intrinsic nature and beingness. In condoning the alteration of an animal's intrinsic nature through bio-engineering so that it will be happy under unnatural and sub-optimal conditions, a benevolent form of anthropocentrism takes root. However, from a non-anthropocentric perspective (and a non-anthropomorphic one that speaks not of the 'happiness' of animals but of letting them be) Rollin's proposal of eugenics for animals is not acceptable. This leads us to a more specific discussion of the application of transgenic manipulations for the purported benefit of animals. I shall argue below that most of these claims have little or no validity.

Animal Welfare

A common assertion by animal production technologists is that genetic engineering is simply an extension of the age-old practices of selective breeding and cross-breeding (or hybridisation). Yet ethically and scientifically speaking, these are not valid historical precedents, for reason informs us that genetic engineering is an entirely different process. Whatever analogy exists between the old practices and the new is shattered by the fact that in traditional breeding practices genes cannot be exchanged between unrelated species, whereas, in many transgenic manipulations, they can.

When we compare the wolf with various pedigree dogs we find that selective breeding has caused a host of disorders in pedigree dogs. Given that selective breeding can result in animal sickness and suffering, then genetic engineering is also likely to cause harm to animals. For instance altering the normal size of an animal through selective breeding can result in a variety of health problems, especially metabolic and orthopaedic disorders. Dr Wayne Riser (1985) has demonstrated elegantly how the incidence, susceptibility to, and severity of orthopaedic diseases increases as size and shape either increase or decrease from the ancestral size and shape. Similar problems are likely to arise in animals whose size and physiology are altered through genetic engineering.

Genetic engineering has already caused animal suffering. As more animals are subjected to transgenic intervention (or genetic reprogramming) the probability of animal suffering increases. Below I specify the types and sources of suffering the genetic manipulation of animals can cause.

Developmental Abnormalities
Certain types of genetic engineering involve the insertion of foreign genes into animal embryos. Following gene insertion into embryos, the embryos often fail to develop and are aborted [see 'Discussion I'], because gene insertion techniques are still far from perfect and are often 'hit and miss' (Palmiter and Brinster 1986). Not only do the inserted genes often fail to get into the target cells, they often finish up in cells of the wrong organs. Some embryos which develop abnormally may die *in utero* and be aborted or reabsorbed; others may be born with a variety of developmental defects, some of which are attributable to so-called insertional mutations, or be infertile. Health problems may not be manifested until later in life. Lacking the controls of other 'regulator' genes, some foreign inserted genes produce too much of a certain protein, such as insulin or growth hormone, causing sickness and even death in the experimental animal (Seldon *et al.* 1987). Because of the nature of some forms of genetic reprogramming, there can be no safeguards to prevent animal suffering. These problems are to be expected in the initial phase of creating transgenic animals and in other types of genetic manipulation experiments on animals.

Deleterious Pleiotropic Effects
The term deleterious pleiotropic effects refers to multiple harmful effects of one or more genes on an animal's phenotype – its entire physical, biochemical and physiological make-up. Once the antici-

pated genetic changes have been accomplished through genetic engineering, and the new animal prototypes developed as foundation breeding stock, additional deleterious effects are to be anticipated. These deleterious pleiotropic effects have been shown to occur in transgenic animals (Pinkert 1988; Simons *et al.* 1987a). One example is the many health problems of the United States Department of Agriculture's (USDA's) transgenic pigs [also known as the 'Beltsville pigs'] that carry the human growth hormone gene. These problems were unexpected, since mice and rabbits genetically reprogrammed with the human growth hormone gene did not manifest deleterious pleiotropic effects to anywhere near the same degree. These pigs were arthritic and lethargic and had defective vision arising from abnormal skull growth. Moreover they did not grow twice as big nor twice as fast which was the result when the human growth hormone gene was inserted into mice. The 'Beltsville pigs' had high mortality rates and were especially prone to pneumonia, with the implication that the transgenic manipulation had seriously impaired their immune systems.

The 'Beltsville pigs' demonstrate that pervasive suffering can arise from transgenic manipulation. They also demonstrate that, just because a particular genetic change causes little apparent sickness and suffering in one species, as in the case of transgenic mice, this does not mean that the same genetic change in another species will have the same consequences. In other words, predictions and assurances about the safety and humaneness of genetic engineering cannot be generalised from one animal species to another.

New Health Problems, Disease Resistance and Animal Suffering
It is to be expected that the genetic reprogramming of animals will often cause them illness and death. Existing veterinary knowledge is unlikely to be adequate to deal with the special requirements of animals subjected to genetic reprogramming, therefore additional research will be needed to correct the ensuing health problems and associated suffering.

Researchers contend that, through genetic engineering, animals can be made disease resistant, and that this will help to reduce animal suffering. However, the notion that genetically engineered disease resistance will reduce animal suffering is naive because it reflects a single-cause – microbial – conception of disease. Simply endowing an animal with resistance to a particular strain of bacterium or virus, for example, will not protect it from stress factors, such as overcrowding and confinement in 'factory farm' buildings and the contingent suffering, that make it susceptible to disease in the first place. These conditions are in themselves pathogenic (Fox 1984).

Genetically Engineered 'Bio-boosters'
It is unlikely that new transgenic varieties of farm animals will become part of the food production system within the next five to ten years, with the exception of genetically engineered fish, such as salmon and catfish. Instead, genetically engineered bacteria are being utilised for the cloning and mass production of injectable 'bio-boosters' designed to increase disease resistance, growth rate, milk production, leanness of carcass, fertility and fecundity. The social, economic and ecological consequences and the animal welfare implications of the improper use of these biologics (biological drugs) by farmers is being hotly debated (see Kronfeld 1988). Their desirability is also being questioned in these times of overproduction of meat and dairy produce in the USA and the European Community (EC). The resulting food surpluses cost the public billions of dollars in storage costs, and, coupled with price subsidies to other sectors of the livestock industry, cast a dark shadow over these innovations, which some believe now threaten the structure and future of agriculture. Suffice it to say that 'agribiotechnology' will fail in the long-term if it is not integrated with a sustainable, regenerative agriculture, and is instead used to boost what is fundamentally an ecologically and economically unsound meat-based agricultural system.

Productivity and Suffering
The use of transgenic manipulations to increase the productivity and 'efficiency' (for example growth rate, milk or egg yield) of farm animals kept under intensive husbandry conditions will increase the severity and incidence of animal suffering and sickness. It is already extensively documented that farm animals raised under intensive confinement husbandry systems in order to maximise production and efficiency suffer from a variety of so-called 'production-related diseases' (Fox 1984). The argument that if animals are suffering they will not be productive and farmers will not profit is demolished by the fact that animal scientists, in using the term 'production-related diseases', acknowledge that animal sickness and suffering are an unavoidable and an integral part of modern livestock and poultry farming. Using genetic engineering to make animals even more productive and efficient under these conditions, as many animal production technologists are proposing (see, for example, Faras and Muscoplat 1985), will place their overall welfare in greater jeopardy than ever, because the severity and incidence of production-related diseases will be increased if transgenic animals are kept under the same conditions as farm animals are today.

Erroneous 'Benefits' to Animals

Erroneous claims are also made concerning the potential benefits of genetic engineering for the animals themselves. It has been claimed that genetic engineering could be used to help cure animals of genetic disorders. Almost 200 disorders of genetic origin, for example hip dysplasia, have been identified in highly inbred 'purebred' dogs, and there are dozens that afflict other domesticated species. However, it would be a poor investment to attempt to correct genetic disorders in the germline (sex cells) of affected animals since the termination of genetic disorders in this way is unlikely to be profitable, a point which is also relevant to corporate investment and involvement in human germline manipulation.

On closer examination, it is not mere animal rights sentiment or some spiritual or religious belief that leads us to the conclusion that it is wrong to alter an animal's *telos* by genetically engineering changes in its germline. Beyond the moral polemics of right and wrong there is the cold fact that, regardless of any purported benefits to animals from such genetic engineering, there are safer, less invasive and more practical alternatives such as to change husbandry practices, and the deliberate selection against mutations and other inherited anomalies in domestic animals. The best way to prevent the transmission of inherited genetic disorders is simply to cease the practices of in-breeding and breeding from defective animals. Thus from a purely utilitarian perspective the animal's *telos* need not be altered. The only grounds for doing so would be, at best, as a 'last resort' to save some endangered species whose gene pool is contaminated by a lethal gene.

In sum, the sanctity of being and the inviolability of the *telos* of animals are ethical principles that need not be dismissed for quasi-eugenic reasons that favour genetic engineering. There are many alternative ways to enhance the welfare and overall wellbeing of animals without having to resort to genetic engineering. Furthermore it would be bad medicine to resort to the latter without a thorough consideration of viable alternatives that uphold the ethical principle of the inviolability of an animal's *telos*.

Future Concerns

Genetic engineering and other new biotechnologies, such as embryo transplantation, cloning and the creation of chimeras like the 'geep' – a combination of the genetic material of a sheep and a goat – are relatively recent developments. This means that, at present, there is a total lack of evidence that the welfare of animals subjected to these manipulations can be guaranteed, and that coincidental and contingent suffering to these animals will be avoided. It is wrong to presume or promise that as a result of the application of these new technologies the welfare of animals will not be placed in jeopardy, and that 'unnecessary' suffering will be avoided.

We have clear evidence of suffering in genetically engineered animals whose welfare has been jeopardised by this new technology. Sometimes the birth defects and cancer caused by transgenic engineering of animals were anticipated (see Brinster *et al.* 1984; Woychik *et al.* 1985; Leder *et al.* 1986). On other occasions the animal suffering was not anticipated by the researchers. Diabetic mice, for example, temporarily cured by the insertion of human insulin genes, died later from excessive insulin production (Seldon *et al.* 1987). Federal animal welfare regulations in the USA contain no reference to genetically engineered animals. They relate to the care of animals and are not applicable to the prevention and alleviation of animal suffering following genetic reprogramming (see Box 5.1). Basic guidelines are needed to prevent transgenic animals from unnecessary suffering. Furthermore most species used in transgenic research, including rodents and farm animals, are excluded from the protection of the Federal Animal Welfare Act (AWA), a situation which I believe demands immediate attention.

Box 5.1:
Federal Protection of Genetically Engineered Animals in the USA

It is noteworthy that the two major groups of animals utilised in genetic engineering studies – livestock, and mice and rats – are not currently covered by the Animal Welfare Act (AWA). Yet proponents of genetic engineering claim that sufficient animal welfare protection exists for all animals used in biomedical research in the USA. Current deficiencies in the AWA which prevent even minimal protection of animals produced and utilised by genetic engineering methods include the following:

Lack of Coverage for Farm Animals
The term 'animal' as defined the AWA (P.L. 99–198) specifically excludes farm animals if those animals are used in agricultural research (NAVS 1988). The AWA does cover farm animals used in biomedical research, but the Veterinary Services of the Animal and Plant Health Inspection Service (APHIS) of the United States Department of Agriculture (USDA), which administers the AWA, decided administratively not to enforce AWA coverage for biomedically utilised livestock.

Livestock are being used widely in genetic engineering studies at present, and increased utilisation of livestock species is anticipated

---

Box 5.1 continued ...

in the future. Therefore farm animals which are subjected to genetic engineering, not only to increase productivity for food and fibre purposes but also to produce various biologics and pharmaceuticals, should be covered under the AWA. Furthermore farm animals used in biomedical research should be inspected by the APHIS, and the protection afforded by the AWA should actively be enforced by the USDA.

Lack of Coverage for Mice and Rats
Mice and rats are covered under the AWA because by definition they are 'warm-blooded animals', but they are specifically exempted from coverage under the APHIS regulations and are not currently being inspected. Mice and rats are now being used extensively in genetic engineering research, and coverage for these species should be specifically included in the AWA definition of 'animal'.

The National Institutes of Health (NIH) Guidelines (P.L. 99–158) do cover livestock, mice and rats and have standards for them. However these standards apply only to animal research sponsored by NIH funds, and some genetic engineering research does not fall into this category. Furthermore the NIH does not routinely monitor compliance with their guidelines by on-site inspection. Comprehensive 'letters of assurances' from the universities or research centres receiving NIH funding are required, and in these letters researchers must give details of their animal care procedures and facility layouts, but no on-site inspections occur unless unusual circumstances prevail. The NIH expects compliance, but has no enforcement powers except the threat of withdrawal of NIH funds. Enforcement of animal welfare standards for laboratory animals is left to the APHIS, which, as previously stated, excludes mice, rats and livestock from AWA coverage.

Information supplied by Dr Nancy E. Wiswall, Research Associate, The Humane Society of the United States

Genetic Parasitism?

Mice have been transgenically manipulated so that they secrete human tissue plasminogen activator (TPA) in their milk. TPA helps to remove blood clots from heart attack victims. Cows and sheep may be the next animals to be used for producing many useful pharmaceuticals in their milk, such as human blood-clotting Factor IX for haemophiliacs [see Chapter 3] and alpha-1-antitrypsin for emphysema victims (Wagner 1985; Choo *et al.* 1987). If there is no animal suffering following certain transgenic manipulations, then it is difficult to argue against such 'molecular farming' where animals are used as 'protein factories'. After all, they have long been exploited for products far less vital to human health, such as meat, hides, milk and eggs. Yet does a history of exploitation establish an ethically valid precedent for continued and intensified exploitation?

It may be that drugs manufactured by transgenic cows will be more effective and cheaper than the same drugs manufactured by genetically engineered bacteria. The use of transgenic animals as 'protein factories' for drug manufacturing should be accepted only on condition that the animals are kept under highly humane husbandry conditions that fully satisfy their behavioural and social needs. In providing us with life-saving pharmaceuticals, we would surely owe them no less. The fact that these animals would be so valuable should help to ensure that they would be well cared for, but rather than keeping them under sterile conditions, their environment should provide for their social and behavioural needs.

Molecular farming – the incorporation of human genes into non-human beings for human benefit – is seen as a form of parasitism: genetic parasitism. It is as abhorrent to some people as the practice of grafting pig livers and chimpanzee hearts into humans. It is difficult for others who do not share such feelings to discover the intuitive wisdom behind this emotional response. This intuitive wisdom informs us that the genetic deterioration of *Homo sapiens* (the human race) will continue and increase in severity as long as the major focus of biomedical research is on developing ways to isolate the defective human genes that produce deficient proteins (for example enzymes and hormones) so these genes can be inserted into bacteria and farm animals to manufacture these proteins for us, rather than developing ways to prevent the transmission of genetic disorders. In sum, molecular farming profits by, and could indirectly contribute to, the genetic deterioration of the human race, if not to medical nemesis *per se* (Leder *et al.* 1986). Such genetic parasitism should be beneath our dignity but, after all, we eat pigs and graft their organs into our bodies.

Environmental Concerns

The impact of genetically engineered animals, plants and microbes on wildlife and wildlands worldwide will be devastating unless genetic engineering is endowed with an environmentally sound philosophy. The genetic modifications of farm animals will enable them to adapt to habitats which are presently unsuitable for them, for example tropical, arid or wetland areas. They will be made resistant to various diseases, for example 'sleeping-sickness', to which they are currently susceptible. Genetically engineered farm animals could escalate the rate of destruction of wildlife habitat, accelerate the decline in bio-diversity and, some predict, lead to the end of the natural world. This final transformation of remaining wilderness areas and the reduction in bio-diversity worldwide by the introduction of animals genetically adapted to new habitats will be intensified further by the introduction of genetically engineered plants likewise adapted to new habitats.

The risks to wildlife populations and global bio-diversity may far outweigh any short-term benefits that recombinant DNA technology may promise, but this need not be the case. The biotechnology industry has the power and responsibility to initiate ecologically appropriate programmes and policies, and to act creatively for the sake of the future of all life on earth. This entails not simply respecting the natural world or the sanctity of being, but realising that it is prudent economically in the long-term to preserve the integrity and diversity of biotic (living) communities.

Animal Patenting in the USA

In the final analysis, the public interest and good of society will be served by genetic engineering only if the long-term ethical, social and environmental consequences of this new technology are carefully considered.

The genetic engineering and patenting of life-forms (see Box 5.2) reflect an exploitative and 'dominionistic' attitude toward living beings that denies any recognition of their inherent nature. As one veterinarian and biotechnologist, Dr Thomas Wagner, has stated publicly, a cow is 'nothing but cells on the hoof' (Choo *et al.* 1987). From an ethical perspective, this objectification and 'mechanomorphic' perception of animals reflects a cultural attitude that is contrary to the concept of the sanctity of being and the inherent value of other living creatures.

Financial incentives behind the genetic engineering of animals, spurred on by the patenting of transgenic animals, will add to the suffering of animals under our dominion if the momentum of research and development in this field is not slowed down by placing a moratorium on the patenting of animals, and by encouraging public debate and participation in the decision-making and policies of industry and government. Much tighter and more appropriate animal welfare regulations and better guarantees of corporate responsibility with regard to the wellbeing of animals are needed to prevent the suffering that will otherwise be caused to animals subjected to certain forms of genetic reprogramming purely for reasons of utility (Wagner 1985).

Box 5.2:
Animal Patenting in the USA: Synopsis of the Effort of a Public Pressure Group Coalition to Place a Moratorium on the Patenting of Genetically Engineered Animals

Background
In April 1987 the US Patent and Trademark Office (PTO) announced that patents could be issued on genetically engineered animals. Subsequently the US Senate passed Senator Mark Hatfield's (Republican, Oregon) amendment to an appropriations bill. This amendment would have temporarily blocked such patenting, but the House of Representatives declined to accept the Hatfield amendment in conference on the bill since the PTO commissioner, Donald Quigg, stated that no such patent would be issued until after March 1988. In April 1988 the PTO issued the first patent on a mammal to Harvard University and Du Pont (NAVS 1988). Although the subject is a mouse, the patent covers any mammal containing a human gene which causes breast cancer.

In June 1987, the Foundation on Economic Trends together with the Humane Society of the United States initiated the formation of a coalition of animal welfare organisations, environmental pressure groups, farm groups and religious organisations to oppose animal patenting. The concerns of this coalition are the long-term ethical, animal welfare, environmental, economic and governmental consequences of the patenting of animals.

The House Judiciary Subcommittee on Courts, Civil Liberties and the Administration of Justice held a series of four hearings on the issue. John Hoyt, President of the Humane Society of the United States, testified at the June 1987 hearing. No hearings were held on the Senate side.

In August 1987 Congressman Charlie Rose (Democrat, North Carolina) introduced a bill (House of Representatives Bill 3119) into the House of Representatives which would have placed a moratorium on the patenting of genetically engineered animals for two years, and Senator Mark Hatfield introduced an open-

Box 5.2 continued ...

ended moratorium bill (Senate Bill 2111) into the Senate in February 1988. These bills were supported by the Humane Society of the United States, and the public pressure group coalition. Sixty members of the House of Representatives co-sponsored Congressman Rose's bill (HR 3119), including a number of House Judiciary Committee members, but the bill was vehemently opposed by the biotechnology industry and the Reagan Administration.

Some members of the Judiciary Courts Subcommittee were receptive to arguments that time is needed to study the potential problems of the patenting of animals. However, the majority of members of the Judiciary Courts Subcommittee and the full Judiciary Committee appeared to be more concerned that a moratorium would retard the advantage that the US biotechnology industry currently holds over the rest of the world. Therefore Congressman Bruce Morrison's (Democrat, Connecticut) compromise effort, which would have prevented animal patenting until the appropriate regulatory system was in place, was defeated in both the Judiciary Courts Subcommittee and the House Judiciary Committee. The Humane Society of the United States supported the Morrison effort. Instead, the Judiciary Courts Subcommittee approved the Kastenmeier (Democrat, Wisconsin) bill (HR 4970) which would exempt researchers and small farmers from patent liability and prohibit the patenting of human beings.

In a later committee an amendment proposed by Synar (Democrat, Oklahoma), which expands the farmer exemption to include larger farmers, was approved by voice vote. Also by voice vote, the committee passed an amendment by Moorhead (Republican, California) deleting the research exemption. Kastenmeier's bill (HR 4970) was passed by the House of Representatives in September 1988.

Current Status
By September 1988 the Kastenmeier bill (HR 4970) had been referred to the Senate Judiciary Patents Subcommittee (DeConcini, Democrat, Arizona).

The Humane Society of the United States would prefer a moratorium on the patenting of animals but it did not have enough support in Congress in 1988. The public pressure group coalition supports Kastenmeier's bill (HR 4970) because it would provide a deterrent to farm animal patenting. However, the coalition would

Box 5.2 continued ...

prefer to see amendments which reinstate the research exemption and which clarify the human being exemption so that it includes foetuses.

The biotechnology industry has said it is now opposed to the Kastenmeier bill (HR 4970) because of the expanded farmer exemption. Towards the end of 1988 the industry announced opposition to any farmer exemption whatsoever.

Outlook
The incentive for passage of the Kastenmeier bill (HR 4970) is that the moratorium issue would be laid to rest. Passage of such a bill currently hinges on a compromise between the biotechnology industry and the farm groups. However, if no bill is passed, and the PTO continues to process patents, the coalition's argument for a moratorium becomes stronger. (There are over 40 animal patent applications pending.) The issue is to be reconsidered when Congress reconvenes in 1989.

Information supplied by Martha Cole Glenn, The Humane Society of the United States, September 1988

Conclusions

We should take advantage of the hindsight that the history of the evolution of industrial society offers us. Those who have a vision of some future Utopia to be brought about by genetic engineering – and there are many, evidenced by the fact this is the most outstanding growth industry of the 1980s – may well be suffering from historical amnesia, or what theologian Father Thomas Berry calls 'technological entrancement' (see Illich 1977). As he sees it, rather than 'reinventing' ourselves to assume a more planetary role, this entrancement leads us to recreate the world in our own image to serve our own needs, no matter how spurious.

History teaches us that the consequences of 'technological entrancement' have been highly destructive to other living creatures, to ourselves and to the environment. The problem remains, however, that if this newly acquired power over the genes of life, like our power over the atom, is not exercised with the wisdom and humility of a planetary stewardship, with compassion for all living beings, and respect for the ecological interdependence within the created order, the cost to all will far outweigh the benefits to the few.

Discussion I

Baroness Edmi di Pauli: Dr Fox, do you consider that the increase in neuroses in humans has anything to do with our consumption of animals which have led an unhappy life, suffering the mental and emotional effects of 'factory farming'?

Dr Michael Fox, HSUS: There is an old saying of the Sioux Indians that when a young man is out hunting he should endeavour to kill the deer swiftly with one arrow, otherwise he will feed its fear to his family. I think the real issue here is probably related to the lactic acid and other changes in the muscles of the factory-farmed animals that we eat. I believe there is something in the folk wisdom of these traditional societies that we could benefit by reflecting upon.

Anon: Professor Webster, do you think that the excesses of 'factory farming' together with the excesses of genetic engineering will turn public opinion against using animals to produce our food?

Professor John Webster, University of Bristol: If we should ever reach the point where farm animals were no longer working with us to produce our food then they would become our competitors and the prospects for animal abuse would become much greater. If you neglect animals you commit sins of omission. It is certainly true that ill-health is one of the major problems of animal welfare at the moment and I am concerned about the somewhat Victorian conditions of squalor and poverty under which some farm animals are kept. Under these circumstances, improvements in husbandry and hygiene have a far greater potential for improving farm animal welfare than breakthroughs in genetic engineering.

Alan Long, Vegetarian Society: Dr Murphy, could you explain why sheep are being used to produce human proteins rather than bacteria, as in the case of insulin?

Dr Caroline Murphy, RSPCA: The problem with producing long mammalian proteins in bacteria is the length of the proteins and the length of the gene that codes for them. Unlike mammals, bacteria do not have big chromosomes; they only have little rings of DNA. The best way to produce long mammalian proteins is to use mammalian cells. However, there is even a limit to the length of DNA which can be inserted into mammalian cells, which is why Factor IX is being made by transgenic sheep rather than Factor VIII. Factor VIII is a more useful protein, but it is a longer protein. It is possible that instead of transgenic mammals we will be able to use yeast cells to produce mammalian proteins, and this would be preferable from an animal welfare point of view. But if it proves impossible to use yeast cells then we will need an alternative source of human blood protein other than extracting them from human blood, because the risk of contamination with harmful viruses, such as HIV, is too high. I was shocked to hear the Director of the Haemophilia Centre in London recently report that parents no longer discuss whether or not to abort haemophiliac foetuses; they are simply not prepared to continue with the pregnancy because of the risk that the child will be given contaminated blood coagulants.

Dr Peter Wheale, Bio-Information (International) Limited: My understanding from Professor Mark Williamson of the Advisory Committee on Genetic Manipulation (ACGM) Planned Release Sub-Committee is that transgenic sheep should be penned at all times. I was therefore surprised to see Dr Bulfield's slide depicting transgenic sheep apparently in open fields. Would you care to comment on this?

Dr Grahame Bulfield, AFRC: At the AFRC Research Station in Edinburgh, we are following the draft ACGM Guidelines on Transgenic Animals [final version published in January 1989, ACGM/HSE/Note 9]. These draft guidelines discuss penning and make recommendations for types of fencing along similar lines to those of the Ministry of Agriculture, Fisheries and Foods (MAFF). The sheep in the picture have not been 'released' and are in a paddock which meets with the requirements of the draft guidelines.

Anon: Dr Murphy, what kind of regulations exist for controlling the transgenic manipulation of animals in the UK?

Dr Caroline Murphy, RSPCA: In the UK transgenic animals

have to be kept in Home Office inspected premises. However, on the continent of Europe there is no such provision. One consequence of this is that on the continent a transgenic pig was kept on a commercial pig farm. When the researchers who had engineered the pig and were studying its biochemistry returned to work one Monday morning, they could not find it. In fact, the pig had died of pneumonia over the weekend and the carcass had been disposed of. This could not have happened in the UK. The provisions of the Animals (Scientific Procedures) Act (1986) ensure that research animals are given the highest standards of animal husbandry and are kept on licensed premises. There is some overlap between the provisions of the Animals (Scientific Procedures) Act (1986), which is administered by the Home Office, and the genetic manipulation guidelines administered by the ACGM of the Health and Safety Executive (HSE) [see Chapter 23]. We need to ensure that these two agencies work together in harmony.

Patricia Spallone, feminist writer: In discussions on genetic engineering and embryo research there is a tendency to use gender neutral language which encourages us to overlook the fact that in this research females are the experimental subjects. It is they who go through the medical and scientific manipulations – they are super-ovulated with fertility drugs and hormones, their eggs are extracted and embryos are transferred into their wombs.

Anon: Surely, Dr Bulfield, enormous suffering must have been caused to the transgenic mouse which carried the abnormally large transgenic mice in her womb. Also, surely giving birth to these 'super mice' would have been an agonising process.

Dr Grahame Bulfield, AFRC: Animals are made large in a variety of ways of which genetic manipulation is one, and classical animal breeding is another. I think that you are right to focus on pain. If pain and suffering is caused to an animal, then it is wrong, irrespective of whether or not genetic engineering has caused it. I think you are focusing on the techniques of genetic engineering rather than on the results of the technique. In the case of the transgenic mice, I have not read in the literature that their surrogate mothers have any problems in giving birth to them.

Ruth McNally, Bio-Information (International) Limited: In his presentation, Dr Bulfield stressed the idea that the new techniques of genetic engineering are different to traditional breeding techniques because they are precise. In response to this assertion, I wish to describe an experiment undertaken in 1987 [see Anderson 1987], which is very similar to the sorts of experiments which Dr Bulfield described. In this experiment 320 mouse eggs were each injected with 200 copies of a foreign gene. Only one mouse was subsequently born that was able to make the foreign protein. Many of the eggs that were injected failed to develop properly because the insertion of the foreign genes into the eggs disrupted the normal development of the mice. Even in those cases where the insertion itself was not detrimental to the development of the mice, the expression of the foreign gene in uncontrolled quantities and in inappropriate tissues had harmful effects upon them [see Wheale and McNally 1988a, pp. 164, 214–5]. Similarly, in the case of human gene therapy – the insertion of foreign genes into human cells – for which there is a protocol being developed in the USA, it has been predicted that one patient in five may succumb to leukaemia as a result of what is called 'insertional mutagenesis' – the disruption of the function of the host cells by the integration of the foreign genes [Wheale and McNally 1988b]. The point about the 'Beltsville pigs' is that the scientists who performed the genetic engineering experiments did not predict that inserting a growth hormone gene would have such severe skeletal effects, that it would affect the animals' skin, and that it would affect their vision [Wheale and McNally 1988a, p. 164]. In short, as Professor Webster suggested, genetic engineering is not so much a science, but a series of novel techniques whose effects are being investigated by trial and error, often at the expense of animal welfare, consumers and the environment.

John Parkin, FAWC: Dr Fox, I would like to ask you a couple of questions about the patenting of animals. The first question is, could not patenting be used as a means of regulating the production of genetically engineered animals? The second question is, in the absence of patenting, would there not be a danger that the companies which developed a 'super pig' or a 'super chicken', for example, would be very reluctant to release this product? Then it might be grown in their own factories under conditions of secrecy which could be worse for the animal's welfare than if information about it were publicly available through patenting disclosure.

Dr Michael Fox, HSUS: Yes, I do think that patenting could be used to regulate the genetic manipulation of animals and that animal welfare could suffer in the absence of patenting law. Were the Trade Secrets Act to be used as an alternative to patenting, then we would never know what was being done to the animals.

However, by the time a patent is disclosed it is too late to protect the animal because the work has already been done. I oppose the patenting of animals for two reasons. First, from the ethical position which opposes commercialisation of ownership of living creatures. And second, to slow down the momentum of the genetic engineering of animals in order to allow more time for a public debate on the whole issue.

References I

ACGM/HSE/Note 9 (1989) *Guidelines on Work with Transgenic Animals* (London: ACGM).

Anderson, I. (1987) 'New genes cure a shivering mouse', *New Scientist*, 5 March, p. 24.

APC (1989) 'Report of the Animal Procedures Committee for 1988', *House of Commons Paper 458* (London: HMSO).

Berg D.E. and Howe, M.M. (eds) (1989) *Mobile DNA* (Washington DC: American Society for Microbiology).

Berry, T. (1978) *The New Story* (New York: Teilhard Association for the Future of Man).

Brem, G. *et al.* (1986) 'Production of transgenic mice, rabbits and pigs by microinjection', *Theriogenology*, vol. 25, p. 143.

Brinster, R.L. *et al.* (1984) 'Transgenic mice harboring SV40 T-antigen genes develop characteristic brain tumors', *Cell*, vol. 37 (2), pp. 367–79.

Carpenter, E. (1980) *Animals and Ethics* (London: Watkins).

Choo, K.H. *et al.* (1987) 'Expression of active human blood clotting factor IX in transgenic mice: use of a cDNA with complete RNA sequence', *Nucleic Acids Research*, vol. 15(3), pp. 871–84.

Clark, A.J. *et al.* (1989) 'Expression of human anti-hemophilic factor IX in the milk of transgenic sheep', *Bio/Technology*, vol. 7, pp. 487–92.

Coles, P. (1989) 'Taking advice from experts', *Nature*, vol. 340, p. 178.

Day, M.J. (1982) *Plasmids* (London: Arnold).

Dickman, S. (1989) 'Oncomouse seeks European protection', *Nature*, vol. 340, p. 85.

DoE (1989) *Proposals for Additional Legislations on the Intentional Release of Genetically Manipulated Organisms* (London: DoE).

Doolittle, W.F. (1982) 'Selfish genes, the phenotype paradigm and genome evolution', in J. Maynard Smith (ed.), *Evolution Now: A Century After Darwin* (London: Nature, Macmillan).

European Commission (1988) *Proposal for a Council Directive on the Legal Protection of Biotechnological Inventions* (Brussels: COM (88) 496 final – SYN 159).

Ezzell, C. (1988) 'First ever animal patent issued in United States', *Nature*, vol. 332, p. 668.

Ezzell, C. (1989) 'Transgenic sticky issues', *Nature*, vol. 338, p. 366.

Falconer, D.S. (1973) 'Replicated selection for body weight in mice', *Genetical Research*, vol. 22, pp. 291–321.

Faras, A.J. and Muscoplat, C.C. (1985) 'The impact of genetic engineering on animal health and production', *Journal of Animal Science*, vol. 61 (2), pp. 144–53.

51

FAWC (1988) *Report on Priorities in Animal Welfare Research and Development* (Tolworth: FAWC).

Fedoroff, N.V. (1984) 'Transposable genetic elements in maize', *Scientific American*, vol. 250, no. 6, pp. 65–75.

Fox, M.W. (1984) *Farm Animals: Husbandry, Behavior and Veterinary Practice* (Baltimore: University Park Press).

Fox, M.W. (1987) 'Keeping the lid on Pandora's Box', *Business and Society Review*, Summer, pp. 50–4.

Goodnow, C.C. *et al.* (1988) 'Altered immunoglobulin expression and functional silencing of self-reactive B lymphocytes in transgenic mice', *Nature*, vol. 334, pp. 676–82.

Hammer, R.E. *et al.* (1985) 'Production of transgenic rabbits, sheep and pigs by microinjection', *Nature*, vol. 315, pp. 680–3.

Hansard (1985), *House of Lords Reports*, vol. 469 (23), col. 765, 17 December.

Hansard (1989) 'Ban of useless animal experiments', 839–41, Bill 168, 27 June.

HSC (1987) *Review of the Health and Safety (Genetic Manipulation) Regulations 1978* (London: HSE).

Illich, I. (1977) *Medical Nemesis: The Expropriation of Health* (New York: Bantam).

Joyce, C. (1988) 'Patent on mouse breaks new ground', *New Scientist*, 21 April.

Keller, E.F. (1983) *A Feeling for the Organism* (New York: Freeman).

Kronfeld, D.S. (1988) 'Biologic and economic risks associated with the use of bovine somatotropins', *Journal of American Veterinary Medical Association*, vol. 192, pp. 1,693–96.

Leder, P. *et al.* (1983) 'Translocations among antibody genes in human cancer', *Science*, vol. 222, pp. 765–71.

Leder, P. *et al.* (1986) 'Consequences of widespread deregulation of the c-myc gene in transgenic mice: multiple neoplasms and normal development', *Cell*, vol. 45 (4), pp. 485–95.

Lichtenstein, C. (1987) 'Bacteria conjugate with plants', *Nature*, vol. 328, pp. 108–9.

McClintock, B. (1978) 'Mechanisms that rapidly reorganise the genome', *Stadler Genetic Symposium*, vol. 10, pp. 25–48.

McClintock, B. (1984) 'The significance of responses of the genome to challenge', *Science*, vol. 226, pp. 792–801.

McGourty, C. (1989a) 'New regulations proposed', *Nature*, vol. 340, p. 88.

McGourty, C. (1989b) 'About-turn on regulations', *Nature*, vol. 341, p. 6.

Mahon, K.A. *et al.* (1987) 'Oncogenesis of the lens in transgenic mice', *Science*, vol. 235, pp. 1,622–8.

Nature (1982) vol. 300, 16 December, front cover.

NAVS (1988) 'Mouse patented', *The Campaigner and Animals' Defender*, March/April, p. 31.

OECD (1986) *Recombinant DNA Safety Considerations: Safety Considerations for Industrial, Agricultural and Environmental Applications of Organisms Derived by Recombinant DNA Techniques* (Paris: OECD).

OECD (1987) 'National Policies and Priorities in Biotechnology', Minutes of Meeting, Toronto, April, Broschure Seite 110.

Osborn, M.J. (1984) Excerpt from a statement sent to the NIH Recombinant DNA Advisory Committee (RAC) in support of transgenic animal research, November.

OSTP (1986) 'Co-ordinated Framework for the Regulation of Biotechnology', *Federal Register*, vol. 51, no. 123, pp. 23,301–50, 26 June.

Palmiter, R.D. *et al.* (1982) 'Dramatic growth of mice that develop from eggs microinjected with metallothionein-growth hormone fusion genes', *Nature*, vol. 300, pp. 611–5.

Palmiter, R.D. and Brinster, R.L. (1986) 'Germ-line transformation of mice', *Annual Review of Genetics*, vol. 20, pp. 465-99.

Perry, M.M. (1988) 'A complete culture system for the chick embryo', *Nature*, vol. 331, pp. 70–2.

Pinkert, C.A. (1988) 'Gene transfer and the production of transgenic livestock', in *Proceedings of the US Animal Health Association* (Columbia, Missouri: US Animal Health Association), pp. 122–34.

Pursel, V.G. *et al.* (1987) 'Progress on gene transfer in farm animals', *Veterinary Immunology and Immunopathology*, vol. 17, pp. 303–12.

RCEP (1989) *The Release of Genetically Engineered Organisms to the Environment* (London: HMSO).

Riser, W.H. (1985) *The Dog. His Varied Biological Makeup and its Relationship to Orthopaedic Diseases* (Mishawaka, Indiana; American Animal Hospital Association).

Rollin, B. (1986) 'On telos and genetic manipulation', *Between the Species*, vol. 2, pp. 88-9.

Rose, M.R. and Doolittle, W.F. (1983) 'Molecular biological mechanisms of speciation', *Science*, vol. 220, pp. 157–62.

Scott, A. (1987) *Pirates of the Cell: The Story of Viruses from Molecule to Microbe*, 2nd edn (Oxford: Basil Blackwell).

Seldon, R.F. *et al.* (1987) 'Regulation of insulin-gene expression', *New England Journal of Medicine*, vol. 317, pp. 1,067–76.

Shapiro, J. (ed.) (1983) *Mobile Genetic Elements* (New York: Cold Spring Harbor Symposium).

Shulman, S. (1989a) 'Cambridge first with local law', *Nature*, vol. 339, p. 496.

Shulman, S. (1989b) 'Cambridge law passes unanimously', *Nature*, vol. 340, p. 88.

Simons, J.P. *et al.* (1987a) 'Efficient production of transgenic sheep', *Bio/Technology*, vol. 6, pp. 171–83.

Simons, J.P. *et al.* (1987b) 'Alteration of the quality of milk by expression of sheep beta lactoglobulin in transgenic mice', *Nature*, vol. 328 (6130), pp. 530–2.

Singer, M. (1983) 'The genetic program of complex organisms', in *Frontiers in Science and Technology: A Report by the Committee on Science, Engineering and Public Policy of the National Academy of Sciences, National Academy of Engineering and Institute of Medicine* (New York: Freeman), pp. 15–44.

Singer, M. (1984) Excerpt from a statement sent to the NIH Recombinant DNA Advisory Committee (RAC) in support of transgenic animal research, November.

Tonegawa, S. *et al.* (1978) 'Organisation of immunoglobulin genes', *Cold Spring Harbor Symposium of Quantitative Biology*, vol. 42, pp. 921–31.

Wagner, T.E. (1985) 'The role of gene transfer in animal agriculture and biotechnology', *Canadian Journal of Animal Science*, vol. 65, pp. 539–52.

Wagner, T. (1987) quotation from *Fortune Magazine*, October, p.80.

Watson, J.D. *et al.* (1983) *Recombinant DNA* (New York and Oxford: Freeman).

Weatherall, D. (1987) 'Molecular and cell biology in clinical medicine: Introduction', *British Medical Journal*, vol. 295, 5 September, pp. 587–9.

Webster, A.J.F. (1988) 'Comparative aspects of the energy exchange', in K.L. Blaxter and I. MacDonald (eds), *Comparative Nutrition* (London: Libbey), pp. 37–53.

Wheale, P.R. and McNally, R. (1986) 'Patent trend analysis: The case of microgenetic engineering', *Futures*, October, pp. 638–57.

Wheale, P.R. and McNally, R. (1988a) *Genetic Engineering: Catastrophe or Utopia?* (Hemel Hempstead: Wheatsheaf; New York: St Martin's Press).

Wheale, P.R. and McNally, R. (1988b) 'Technology assessment of a gene therapy', *Project Appraisal*, vol. 3, no. 4, December, pp. 199–204.

Woychik, R.P. *et al.* (1985) 'An inherited limb deformity created by insertional mutagenesis in a transgenic mouse', *Nature*, vol. 318 (6041), pp. 36–40.

Part II
Genetically Engineered Bovine Somatotropin (BST)

6 Introduction II

Dr PETER WHEALE and RUTH McNALLY

In *Conjectures and Refutations* (1969) the philosopher of science, Karl Popper, contends that scientific theory should be falsifiable and can only be accorded provisional acceptance. According to this view all scientific theories and facts are thus always open to challenge (see also Boyle and Wheale 1986). It is important that experimental procedures and results are subjected to a common evaluation. However, the professional process whereby knowledge is legitimised and scientific information is disseminated is important in determining what is to be considered 'valid' knowledge. The nature of this process means that controversy is not uncommon.

A controversy has arisen over the nature and effects of genetically engineered bovine somatotropin, commonly known by its acronym, BST. This product is one of the first products to emerge from the diffusion of genetic engineering into agricultural production. The somatotropins (or somatotrophins) are hormones which promote general body growth. Indeed, BST is also known as bovine growth hormone (BGH).

The genetically engineered BSTs are produced by inserting BST genes into bacteria which then make BST. Companies which produce genetically engineered BST claim that when this hormone is injected into dairy cows it affects lactation, and can increase milk yield by between 15 and 25 per cent. Since 1985 field trials of genetically engineered BST have been conducted, in several EC countries and in the USA, to test its efficacy as a milk-stimulating drug for dairy cows.

The contributors in this part of the book present their views and supporting evidence on the nature of genetically engineered BST and the effects of administering it to dairy cows. Robert Deakin is Marketing Manager of Monsanto, one of the companies which are currently undertaking BST field trials in the EC and in the USA. Eric Brunner is the author of the London Food Commission (LFC) BST Working Party report, *Bovine Somatotropin: A Product in Search of a Market* (1988a). Dr Wolfgang Goldhorn is Veterinary Director of the State Veterinary Service in West Germany, where BST trials are particularly controversial, and Dr Michael Fox is Vice President of the Humane Society of the United States (HSUS).

The controversy over BST begins with a dispute over the nature of BST itself. In Chapter 7 Deakin maintains that the biological effects of genetically engineered BST are identical to those of the BST produced by the cow herself. In Chapter 8 Brunner charges that the National Office of Animal Health (NOAH) – the animal pharmaceutical industry's pressure group – is inaccurate when it claims in its publicity material that genetically engineered BSTs are identical to BST produced by the cow, because there is a difference in chemical structure between the two types of BST. A further source of controversy has been the naming and classification of BST. The manufacturers have been reluctant to classify it as a hormone, preferring to refer to it as a protein, or BST, rather than bovine growth hormone.

Animal Welfare

The debate on the animal welfare implications of BST-treatment embodies a fundamental difference of opinion on the welfare of the dairy cow under existing 'factory farming' conditions. Compare, for example, Figures 7.1 and 9.1 in the respective chapters of Deakin and Goldhorn depicting the metabolic capability of the dairy cow: where Deakin sees surplus milk-yielding capacity, Goldhorn sees chronic metabolic stress. Deakin, Goldhorn and Fox (Chapters 7, 9 and 10) point out that the dairy industry already controls key aspects of the life of the dairy cow, including where she lives, what she is fed on, which sire she breeds with, how often she is pregnant, and when she is to die. Whereas Deakin considers that in this context the use of BST to enhance her milk yield is just another 'management tool', Fox is critical of the 'instrumental' view of animals which he believes the genetic engineering industry holds, and concludes that the dairy cow is being turned into a 'biomachine'.

Deakin maintains that BST can be administered to dairy cows without detracting from their health, safety or welfare. This view contrasts with those of Goldhorn, Fox and Brunner, each of whom argues that the increased milk demands which BST places on the cow could cause metabolic disorders and reduce her natural resistance against infectious diseases. Goldhorn is concerned that the injections required to administer BST to dairy cows may create painful swellings and distress the animal. Deakin's response to the suggestion that BST-treatment could result in metabolic disorders is to state that other than in research by Kronfeld (1965) there have been no reported cases of ketosis, one of the types of metabolic disorder predicted to occur.

Deakin and Brunner refer to a letter published in Hansard in December 1987. This letter to the Minister of Agriculture, Fisheries and Food was from Professor Sir Richard Harrison, Chairman of the Farm Animal Welfare Council (FAWC), an independent advisory body to the government which was asked to consider the animal welfare implications of administering BST. Deakin reports that, according to Harrison, the FAWC's BST Working Group found no evidence of any welfare problems arising from the use of BST in the short term (Harrison 1987).

Both Deakin and Brunner cite the part of Harrison's letter which states that the FAWC's BST Working Group found areas of uncertainty which could have animal welfare implications, and advises that before the long-term effects of using BST can be properly evaluated, there is a need for further research to gather additional scientific information which would need careful consideration. But whereas Deakin claims that all of the areas of uncertainty have been addressed in more than 50 papers published since the FAWC report, Brunner, Goldhorn and Fox are not satisfied with the research that has been done. Fox maintains that only a few studies on a small number of dairy cows have been conducted on the animal welfare implications of BST-treatment and that these studies have not been for any significant period of their lives. Fox also postulates that research studies are unlikely to examine the effects on the dairy cows' welfare of misuses of BST, for example, its use to hyperstimulate cows which are in poor condition.

Goldhorn expresses the view that the role of the veterinary profession is to act as a mediator between the interests of farmers and those of animals. He questions whether it is sufficient for veterinarians to protect animals from pain, harm and suffering, and believes that they should also safeguard animals' identity and dignity.

Availability of Information

The aim of the BST trials is to gather data on the long-term effects of administering genetically engineered BST to dairy cows. These data are required before full product licences for BST can be applied for. The reader will note that Deakin stresses the availability of information on BST. This is to counter charges about the secrecy of the BST trials and the confidentiality of data from them, which is a point of conflict in this debate.

In order to obtain permission to undertake field trials, Monsanto and the other companies who have developed and are testing genetically engineered BST had to provide evidence that the milk from BST-treated cows was safe for human consumption. Indeed, on the strength of this evidence milk from BST-treated herds involved in the trials is being distributed and sold to consumers in the USA and Britain. However, these data on consumer safety are not available for public scrutiny.

The Ministry of Agriculture, Fisheries and Food (MAFF) issued Animal Test Certificates for the BST trials in Britain under the Medicines Act (1968). This Act ensures the confidentiality of such data and, as Brunner emphasises, the MAFF has refused to disclose the preliminary results of these trials and their locations. Indeed, Professor Richard Lacey of the University of Leeds, who serves on the Veterinary Products Committee which is assessing the BST trials for the MAFF, has complained that he is prevented from speaking out on the human health risks posed by the BST trials because he is bound by the Official Secrets Act and the Medicines Act (1968). It can be argued that since BST is not a veterinary 'medicine' to improve the health of sick animals, it should not be regulated under the Medicines Act (1968) at all.

The credibility of the sources of information used by the proponents and detractors of the case for BST is a recurring feature of this controversy. Brunner charges that the information distributed by the NOAH is inaccurate. Deakin countercharges that the LFC are acting as 'gastronomic terrorists' for suggesting that milk and dairy products from BST-treated cows could be detrimental to consumer health. Deakin questions the evidence cited by Fox and Brunner which suggests a possible relationship between BST and metabolic disease and suggests that the Compassion in World Farming (CIWF) *Dossier on BST* (D'Silva 1988) is a 'misinformation dossier' – a charge denied by Joyce D'Silva in 'Discussion II'.

Consumer Welfare

There is a conflict of opinion on the composition of milk between Deakin, who states that milk from BST-treated cows is not different from the milk from cows which have not been treated with BST, and Brunner and Fox, who cite evidence for differences in fatty acid composition between milk from BST-treated cows and non-treated cows. Deakin denies that any differences which may exist are significant or detrimental to consumers, and states that BST has no impact whatsoever on the consumer of milk or milk products. However, concern about consumer health was expressed when Deakin revealed that Monsanto does not use human subjects in its studies on the effects of BST or on the effects of the consumption of the milk and dairy products from BST-treated cows (see 'Discussion II').

The consensus of opinion appears to be that there are BST residues in milk from BST-treated cows. However, there is disagreement over the significance of this finding. Deakin claims that BST is species-specific (an assertion contested by Joyce D'Silva in 'Discussion II') and is not active in humans, even if injected. On the other hand, Brunner makes claims to the contrary. He also cites evidence that the milk from BST-treated cows contains raised levels of another hormone – insulin-like growth factor (IGF) – which is biologically active in humans; however, in 'Discussion II' this evidence is challenged by Dr Neil Craven of Monsanto.

There is a difference of opinion on the question of whether milk and dairy products from BST-treated herds should be labelled. Brunner reports that a major retail consortium in the UK has requested that the consumers' right to choose whether they want to purchase milk and dairy products from BST-treated cows should be respected. This would necessitate, at the very least, the labelling of milk from BST-treated herds. Deakin's position is unequivocal – milk is milk is milk (see 'Discussion II'): the milk from BST-treated cows is not different and therefore need not be labelled.

The Dairy Industry

All the contributors to this part of the book agree that overproduction is a serious problem of the dairy industry. However, there is disagreement on what the economic effects of genetically engineered BST on the dairy industry will be. Deakin views BST as a 'management tool' which will help to solve the problems of the dairy farmer, which include restrictive quotas and shrinking income. To support his case he cites a study by the Milk Marketing Board (MMB) in the UK which claims that BST is likely to have a relatively small impact on the number of producers by 1995. On the other hand, Goldhorn and Fox maintain that the problems of the dairy industry will be aggravated by the use of BST. Fox warns that BST could lead to the demise of the small family farm which he believes helps to preserve the diversity and flexibility of farming practices. Brunner cites research in which it is estimated that up to half of US dairy farmers could be bankrupted in ten years by the use of BST.

The Future of BST

Deakin believes that BST will be the most thoroughly investigated veterinary pharmaceutical product ever. The licensing of BST is being regarded as a test case for the acceptability of genetically engineered products for agriculture in the EC because it would be the first genetically engineered agricultural product to be licensed in the EC.

Even if it were demonstrated that the safety, quality and efficacy of genetically engineered BST meet the standards required for a product licence, consumers and farmers may still wish to reject it on broader social, economic and animal welfare grounds, that fall outside the evaluation criteria of the licensing authorities. Indeed, pressure is being applied on the European Commission by consumers' and small farmers' groups to broaden the criteria for assessing new veterinary pharmaceuticals so that, in addition to minimum standards of safety, efficacy and quality, such products are also assessed on social and economic grounds.

The dilemma that the European Commission faces is that if genetically engineered BST were to be granted a product licence in the USA, failure to license it in the EC could be construed as a non-tariff barrier to trade and could provoke US trade sanctions against EC imports into the USA. Moreover, in these circumstances transnational pharmaceutical companies may decide to withdraw their research and development activities from the EC (Dickman 1989a).

Faced with conflicting pressures over the licensing of BST as a veterinary product, in September 1989 the EC Agriculture Minister announced a 15-month moratorium on the commercial use of genetically engineered BST. However, this should not be seen as a victory for the anti-BST lobby since it will take the manufacturers of BST at least this long to complete the technical reviews necessary in order to support their applications for product licences for BST (Dickman 1989a, 1989b).

Public opinion and the actions of pressure groups are important influences over the future of BST. Deakin cites the results of the *Omnibus Survey on Milk Consumption* (1988) which found that the reported pattern of milk consumption was similar between those who were aware and those who were unaware of the sale of milk from BST trials. He is confident that once people understand the nature and effects of BST they will accept it.

On the other hand, the CIWF survey in the UK found 82 per cent of consumers surveyed opposed to the use of BST in dairy farming, and Brunner cites a US study which found that a great many consumers might reduce their consumption of milk significantly because of the use of BST. In response to pressure from farm organisations and environmentalist groups, five of the leading supermarket chains in the USA, involving 2,500 supermarkets, assured the public that their brands of dairy products will not contain milk from BST-treated cows (Shulman 1989).

In view of the fact that the social, economic, environmental and animal welfare consequences of using BST in the dairy industry have not been resolved, a majority of those attending the Athene

Trust Conference passed a resolution demanding that the current practice of pooling milk and dairy products from BST-treated cows with the milk and dairy products of untreated cows should be prohibited (see 'Conference Resolutions').

7 BST: The First Commercial Product for Agriculture from Biotechnology

ROBERT DEAKIN

New developments in technology, in such fields as communications, medicine and biotechnology, often out-pace public comprehension. People who do not understand, and indeed some who *refuse* to understand, the developments in biotechnology react by presenting a 'Frankenstein's Farm' image of these developments along the lines of escaped mutants performing in 'B' movie catastrophes, such as 'The Tomato That Ate Tokyo'. These images of the potential threats of biotechnology are complete nonsense. While they may make eye-catching headlines, such images do nothing to improve genuine public understanding. In the nearly 20 years since the first gene was spliced, not one single mishap has occurred anywhere in the world.

Those involved in basic research, for example that which produced the half-sheep/half-goat 'Geep', could and should do much more to explain the importance and relevance of their work, for it is generally public funds that support such research, not the pharmaceutical industry. Nonetheless, I am completely convinced that the fruits of biotechnology research which will finally emerge from today's intensely rigorous regulatory procedures will provide safe and valuable benefits to society.

What is BST?

Bovine somatotropin (BST) is one of the first products of biotechnology for agriculture. BST is a naturally occurring protein which is produced in the pituitary gland of the cow. The existence of BST and its actions have been known for over 50 years. The action of BST in the growing animal is to co-ordinate various growth processes in bone, cartilage, and muscle tissue. In the mature dairy cow, it regulates the production of milk.

Since the bovine species was first domesticated, farmers have been managing the cow. They feed her and try to keep her healthy and happy, but they also govern when and by which sire she is impregnated and many other aspects of her life, including its length. In the not too distant future they will also have at their disposal a new management tool: BST.

BST is a biotechnology product in the sense that biotechnology is used in manufacturing it. The genetic code for BST has been removed from the cow and transferred to a non-pathogenic microbe. The microbes, which multiply under highly controlled conditions, are then killed and the BST is separated, rigorously purified, mixed into a sustained-release formulation, and put into sterile single-use syringes. Essentially the same process is used to produce human insulin in microbes. The BST administered to the cow is biologically identical to that which is produced by the cow herself.

Problems of the Dairy Industry

The problems of agricultural overproduction are a major topic of debate in the European Community (EC). With respect to the dairy industry, it is sufficient to say that the EC policies of the late 1970s and early 1980s encouraged farmers to produce much more milk than the Community could consume or export. Milk processing plants were built solely to produce butter and skimmed milk powder for EC intervention under the Common Agricultural Policy (CAP). Enormous stocks of dairy products resulted which, according to Sir Geoffrey Howe, then the UK Minister for Agriculture, Fisheries and Food, cost the citizens of the EC £240 per cow per year to maintain. The 'eurocrats' responded with quotas and 'superlevies' which have been effective in reducing stocks and surpluses, and have resulted in the greatest socio-economic impact on dairy farming since the EC was formed.

Farmers are now faced with frozen or declining production quotas, and the prices of farm output are mandated to rise at a lower rate than inflation. Their income is being squeezed and they must therefore find ways to reduce their costs. As annual production is reduced to meet annual demand we are seeing a new phenomenon – seasonal shortages. These seasonal shortages are sufficiently significant that the UK Milk Marketing Board (MMB) allotted the whole of 1989's scheduled price increase to a premium price increase of up to 50 per cent for the months of late summer and early autumn in order to encourage milk production at that time.

How Could BST Help?

Figure 7.1 illustrates the lactation cycle of the dairy cow, indicating her changing body weight, feed consumption and lactation curve (milk yield) during the course of the cycle, and also illustrating the effect of BST on the lactation cycle.

A cow lactates (gives milk) for approximately 305 days per year. Starting at calving, her milk yield rises to a peak about 50 days later, and then declines steadily over the remainder of her lactation, after which she is dried-off in preparation for her next calving (see 'normal lactation curve', Figure 7.1).

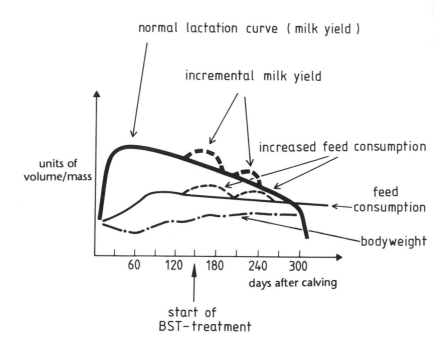

Figure 7.1:
Lactation cycle of the dairy cow with and without BST-treatment

As the lines representing 'feed consumption' and 'body weight' on Figure 7.1 indicate, the cow's feed consumption lags behind her total requirements during the first 70 days of her lactation. During this time she meets the energy demands for milk production by utilising her body fat (Mother Nature intended this to happen) and eating a diet supplemented with concentrated feeds. 70 days or so after calving, her milk yield falls and food consumption increases so that her energy intake meets her energy requirements. The need for concentrated feeds declines and the excess intake is used to restore the cow's body fat (also called her 'condition' or 'reserves') for the next lactation cycle.

BST can be used to best advantage for the farmer and the cow in the latter part of the lactation cycle, from approximately 150 days after calving onward. At this stage in lactation cows tend to have excess feed intake capacity compared to milk output, and can thus at this time produce more milk using lower-cost feeds, such as home grown forage (see 'incremental milk yield' and 'increased feed consumption' in Figure 7.1).

The use of BST in the latter part of the lactation cycle removes the pressure for ever increasing peak yields in early lactation. BST can be used seasonally, such as in late summer when milk prices are higher, and BST can be used by the farmer who, for whatever reason, has fallen behind quota and will lose income unless the rate of production is increased

Numerous studies on the economic effects of using BST in dairy farming have been produced by respected economists throughout the EC (see for example, de Hoop *et al.* 1987; Kalter 1985; Kalter *et al.* 1984; Mouchet 1987). These studies have examined the potential benefits to individual farmers operating under a variety of size, region and farm enterprise situations. BST's benefits arise from its flexibility to fit these various situations.

These studies have also considered the macro or structural impact of BST on various farming regions, individual countries, and the EC as a whole. The conclusions are unanimous: whilst there is an inevitable continuing consolidation in dairy farming, the effects of BST will be negligible. Dairy farm numbers are indeed falling in Europe. They have fallen by almost 40 per cent over the last ten years and will probably fall by about 20 per cent over the next ten years. (Interestingly, true family farms, with 40–80 cows, are forecast to be the most stable type of dairy farm.) To quote the study by Poole (1987) of the UK Milk Marketing Board (MMB): 'From the various scenarios given in this paper (81 were examined), the main result that seems to come out is the relatively small impact that BST is likely to have on the number of producers by 1995.' A similar report by the Dutch Agricultural Economics Institute (LEI) predicted that BST could actually help some farmers to stay in business longer (van der Giessen 1986).

The Health, Safety and Welfare of Cows

One aspect of the intensive public debate on BST which I have found most discouraging has been the frequent and deliberate use by BST opponents of partial information taken out of context, or outright misinformation. It is particularly baffling to me since I would expect that anyone purporting to be defending consumer welfare, animal welfare, or farmer welfare should be driven to seek the truth. I will defend anyone's right to ask questions, but have they the right to repeatedly ignore the answers?

In the Compassion in World Farming (CIWF) 'misinformation' dossier on BST (D'Silva 1988), for example, there is a quotation which starts: 'Few trials have lasted long enough.' This quotation is taken from a paper published before any of the current series of studies had even started. Now, however, a complete bibliography on BST would have entries numbering in the hundreds, including studies now in their fourth year. (The references for some of these papers are included with the references for Part II of the book.)

At the annual meeting of the American Dairy Science Association held in June 1988, over 50 papers on BST from more than 20 institutes were presented. These papers covered the basic mechanisms of the function of BST, clinical trials, acute and chronic toxicity, milk quality and processability. The papers reported on clinical trials which included administering BST to dairy cows in doses of up to 60 times the commercial dose for short periods, and doses six times the commercial dose for up to two lactations. These studies are equivalent to more than double the maximum lifetime treatment with BST that a cow could receive on a dairy farm. Let me emphasise that these are published papers and they are available.

Some critical observers have predicted that BST would result in significant upsets of the cow's metabolic balance. A CIWF quotation from Kronfeld (1965) states, 'Many of the trials have been conducted with very small numbers of cows' (D'Silva 1988). In fact, Kronfeld's 1965 pioneering work on BST included only two cows, and is the only study ever to suggest a possible relationship between BST and metabolic disease.

Now, if you take the trouble to look across all the published reports, you can see that trials of BST have included tens, hundreds, and if pooled, thousands of cows, and they all support the statement that BST has no ill-effects on the cow's health and wellbeing. Those who until recently warned of grave metabolic disorders in BST-supplemented cows should be happy to hear that there has not been one reported case of ketosis, for example, the condition they and Kronfeld most feared and anticipated.

Turning to the cow's welfare, which is in fact difficult to separate from safety and health, one must refer to the BST report of the Farm Animal Welfare Council (FAWC) (see Harrison 1987). The FAWC is a staunchly independent body whose unpaid members include welfarists, farmers, scientists, a lawyer, an auctioneer and several veterinarians. They are the UK government's official reference on the welfare of farm animals and they advise the Minister of

Agriculture, Fisheries and Food on legislative or other changes thought to be necessary.

In 1987 the Ministry of Agriculture, Fisheries and Food (MAFF) asked the FAWC to examine the animal welfare implications of BST. The FAWC held hearings and observed animals under BST-treatment. They concluded that: 'There is no evidence of BST jeopardising treated animals' welfare in the short-term.' They also added that there were still 'areas of uncertainty, which could have welfare implications where additional scientific informed opinion may be required.' The report, which was addressed to John MacGregor, then Minister of Agriculture, Fisheries and Food, and subsequently published in Hansard (the published proceedings of the House of Commons), went on to list seven 'areas of uncertainty' (Harrison 1987).

All of these areas of uncertainty are legitimate areas which must be addressed in any application for licensing of a new veterinary medicine. All of these areas will be addressed in the BST registration dossiers, and in fact all of the issues have been addressed in more than 50 papers published since the FAWC report. The ultimate conclusion is not simply that there is no evidence that BST might jeopardise the welfare of treated cows, it is that there is now a large body of evidence that demonstrates that BST has no detrimental effects on their welfare.

Milk Quality and Consumer Safety

The world's number one convenience food, the humble pint of milk, has come under greater scrutiny than ever before. As a result of the research effort into the use of BST the chemical constituents of milk are now known in greater detail than ever before. For, in order to determine whether BST-supplementation has any effect on milk, one must first thoroughly understand the basic substance.

We (and when I say 'we', I refer to the international scientific community of experts in this field and those others who have done their homework) now know that milk is not just an innocuous white liquid with a few per cent each of fat, protein and lactose, and mostly water. We know that milk contains at least eight types of protein, three types of fats, eight minerals, 50 enzymes, sugar, eight vitamins and 24 hormones, including five steroids and 19 peptides. We know what is in milk, we know how much is in milk, we know by how much the constituents vary in milk from breed to breed, cow to cow, season to season, diet to diet and, of course, with and without BST-supplementation.

We have also conducted trials on the processability of milk. Trials have been conducted making Gouda cheese, Emmental, Brie,

Camembert, Cheddar, Colby, Wensleydale and yoghourt. In study after study the results are the same – BST has no significant effect on milk quality at all.

Let us just review a few facts about BST and consumer safety.

1   Milk from BST-supplemented cows is not different from the milk from cows which have not been treated with BST. Even the amount of BST, which is present in trace quantities in all fresh milk, is not increased by BST-supplementation of the cow.
2   BST is a protein – it is fully digested by people who consume it orally. This statement has been corroborated by Professor A. McNeish, a leading British paediatric endocrinologist, in rendering an opinion with respect to the inability of newborn infants to absorb BST.
3   BST is species-specific. It has no activity in humans, even if injected.

BST has no impact on the consumer of milk or milk products whatsoever. Anyone who implies otherwise, for example the London Food Commission (see Brunner 1988a) [see also Chapter 8], which has selected a single data point from a single, small scale, short-term trial with a non-commercial dose to imply that the fat content of milk will rise by 27 per cent as a result of BST use, is nothing less than a scaremongerer practising gastronomic terrorism in an effort to confuse and frighten consumers and farmers.

I shall now briefly consider consumer perceptions about milk from BST-supplemented cows. Assessing consumer attitudes is a very difficult thing to do and, with new concepts unknown to the respondents, extremely difficult to do without introducing bias. In 1988 CIWF issued a press release indicating 83 per cent of consumers surveyed opposed BST. People surveyed were asked to agree or disagree with the following single statement: 'The daily pinta should remain as it is and not come from cows injected with BST'. No detailed explanation of what BST is, or does, its safety, nor the fact that it does not affect the daily pinta in any way, was provided. It was simply described as a genetically engineered protein hormone which increases the amount of milk which comes from dairy cows.

During 1987 and 1988 BST was covered in more than 1,200 press articles, and given 20 hours of radio and television coverage in the UK. In November 1988 a survey was conducted to see if this inundation of primarily negative media coverage had made any impression on consumers (*Omnibus Survey on Milk Consumption* 1988). Field work and analysis of this survey was done by the British

Market Research Bureau Limited. Face-to-face interviews were conducted amongst a nationally representative sample of 1,908 adults aged 15 years and over. Respondents were questioned on the following issues: their pattern of milk consumption; awareness of BST and the current field trials, and sources of information which they would most trust regarding milk safety and wholesomeness.

The main points to emerge from the results of this study were enlightening. Overall, more people claimed to have increased (24 per cent) than decreased (13 per cent) their milk consumption in the previous 12–18 months. A variety of reasons were given for these changes in milk consumption but, notably, *not a single person* mentioned BST, trials on milk, injections to cows, or other related issues. Of 205 people who had some knowledge of BST, 18 per cent (57 people) were also aware that the milk from BST trials was being sold to the public. However, their reported pattern of milk consumption was similar to those who were unaware of the sale of such milk.

There was little regional variation in the level of awareness of BST and no significant difference in milk consumption, despite a greater level of media coverage in certain regions. When asked without prompting where they would expect to obtain information which they would most trust on milk safety and wholesomeness, four times as many people (27 per cent) said the MMB than any other individual source. The other sources most frequently mentioned were the dairy farmer (7 per cent), the family doctor (6 per cent), the milkman (6 per cent) and the MAFF (6 per cent). After being shown a list of twelve possible sources of information on milk safety and wholesomeness, the MMB (59 per cent) remained the most favoured source. Other sources of trusted information on milk safety and wholesomeness most frequently mentioned included the family doctor (37 per cent), the National Dairy Council (29 per cent), and the MAFF (28 per cent). The London Food Commission – a consumer pressure group which has been actively campaigning against BST – was suggested as a trusted source of information on milk by just one person without prompting, and chosen by less than 4 per cent when offered on a list of possible information sources. The veterinary pharmaceutical industry trade association – the National Office of Animal Health (NOAH) – fared only slightly better.

It should be noted that this survey (*Omnibus Survey on Milk Consumption* 1988) differed markedly in approach from all previous surveys which have attempted to predict consumer attitudes towards BST. The problem with testing attitudes towards a new concept is that you must present a description of the concept to an otherwise unaware respondent. The result is generally a measure of the attitudes towards the presentation of the concept rather than the concept itself. Research indicates that, whether intentional or not, it is very easy to bias a description of BST (especially negatively), and, in fact, very difficult not to.

Early in 1988 Monsanto participated in a series of focus groups, or neighbourhood workshops, organised to exchange views on BST with farmers, veterinarians, doctors, teachers, nurses and consumers. At the end of one such meeting a local National Farmers' Union (NFU) executive, who had organised that meeting, presented two glasses of milk to two women, representatives of the Townswomen's Guild. He explained that one contained milk from a BST-supplemented cow and the other contained 'normal' milk. He asked them to choose which they preferred. Without hesitation both responded, 'We cannot choose, there is no choice, they're both the same.' Such is the response of informed consumers.

Regulatory Process for BST Approval

How can you be sure that what I say is true? How can you be sure that all the evidence that I have quoted exists? How can you be sure that all this research was properly conducted and that the results are thoroughly evaluated? That is where the licensing authorities come in. Dossiers for BST are presently under review by the licensing authorities of the USA, the UK, France and the Committee for Veterinary Medicinal Products (CVMP) of the EC. These licensing authorities are amongst the most rigorous in the world, and by the time that any one company's product is approved, BST will be the most thoroughly scrutinised veterinary drug ever.

Before any EC country can approve BST it must be examined by the CVMP who must render an opinion on it. The CVMP is composed of the heads of the relevant regulatory body in each of the EC member countries. It operates under a specific EC directive, namely, 87/22/EEC, which was established to regulate and coordinate the review of biotechnology and other new technologies for producing human and veterinary medicines.

In the UK the licensing body for new veterinary medicines is the MAFF. The MAFF makes its decision upon consideration of the opinion rendered by the Veterinary Products Committee (VPC). The VPC is a committee of 21 independent experts, including doctors and veterinarians, plus specialists in metabolism, reproduction, residues and other aspects relevant to assessing veterinary medicines. What they assess is laid out in general terms by the Medicines Act of 1968 and they can ask virtually any supplementary questions they deem relevant. The information which the VPC considers when making a decision concerning a new veterinary medicine comes from various sources. For most products, approximately one-quarter comes from company in-house research and the remaining three-quarters from independent investigators at universities and government research institutes. Every test is replicated at different locations.

Some have questioned the independence and even the integrity of the independent investigators. Whilst those making these heinous allegations generally carry no credentials, are self-appointed and are responsible only to themselves, the scientists undertaking the investigations are internationally recognised, subject to peer review, and operate under protocols which, for regulated drugs, require pre-approval by the authorities. They must operate under the rigorous procedures of 'good laboratory practice', and, despite the fact that companies fund the work they do, they are free to publish their findings – positive or negative – as they have, and are doing, with BST.

Conclusions

I shall now review the facts concerning the use of BST in dairy farming.

BST is a product which, when used selectively and intelligently by the dairy farmer, under the supervision of a veterinarian, can help meet the challenges of business – quotas, shrinking income, seasonal factors and unforeseen natural setbacks.

BST can be administered to dairy cows without threatening their health, safety or welfare.

BST has no impact on the consumers of dairy products whatsoever.

BST, being the first of its type – the first such product of biotechnology for the dairy industry – will be the most thoroughly investigated and rigorously assessed veterinary drug ever presented for regulatory approval.

Anything new elicits a certain amount of apprehension. Almost 200 years ago there was a movement against smallpox vaccination as dangerous and unnatural. Artificial insemination was introduced to livestock farming 40 years ago. It too was resisted. But once people understand a new technology and its practical applications they accept it as commonplace.

It is certainly fair to question BST and biotechnology. However, an enquiring mind must be an open mind. As George Bernard Shaw said: 'Progress is impossible without change, and those who cannot change their minds cannot change anything'.

8 Science, Secrecy and BST

ERIC BRUNNER

In 1987 the London Food Commission (LFC) set up the bovine somatotropin (BST) Working Party in response to several press reports drawing public attention to BST. The BST Working Party set out to study all aspects of the proposed use of BST since, at that time, outside the scientific community and industry, virtually nothing was known about it. Membership of the BST Working Party represented 16 consumer, animal rights and environmental groups. Organisations represented on the BST Working Party are listed in Table 8.1.

Table 8.1:
Organisations represented on the London Food Commission BST Working Party

Animal Aid
British Society for Social Responsibility in Science
Compassion in World Farming
Consumers in the European Community Group
Farm & Food Society
Farm Animals Department, RSPCA
Friends of the Earth
Greenpeace
Health Visitors Association
London Food Commission
Maternity Alliance
National Federation of Women's Institutes
National Housewives Association
Soil Association
Transport and General Workers' Union Agricultural Workers Group
Vegetarian Society

A report summarising the findings and proposals of the BST Working Party was published in April 1988 (Brunner 1988a). The report aims to convey to the public answers to a number of questions raised by those on the Working Party who represent many hundreds of thousands of members of organisations. The conclusion of the working party was that the broad socio-economic costs and benefits of the use of BST should be fully assessed before BST is granted a Product Licence by the Ministry of Agriculture, Fisheries and Food (MAFF).

In this chapter I shall consider a number of important questions addressed by the BST Working Party together with the recommendations for information and action which the working party considered necessary to ensure that the public's interests and concerns might be safeguarded.

What is the Nature of BST?

The synthetic BSTs have several different chemical structures. None of these structures is identical to the natural hormone, bovine growth hormone, a protein with 191 amino acid residues. Genetically engineered BST, using the bacterium *Escherichia coli (E. coli)* as a host organism, is synthesised with a single extra amino acid (Monsanto product) or an additional eight amino acids (Elanco product). Assurances have been given that BST hormones pose no risks to human health since they are 'biologically identical' to the natural hormone. This *may* be so, but it cannot be said that they are *chemically* identical.

What will BST be used for?

Biosynthetic versions of bovine growth hormone, known as BST, can boost milk yield by as much as 25 per cent if given regularly to dairy cows. BST also has growth promoting effects in calves and lambs (Johnsson *et al.* 1987; Pell *et al.* 1987).

Some BST trials have been reported in technical publications. In a Monsanto trial, milk yield increased by about 19 per cent with a fortnightly 500mg dose. A trial of BST produced by Elanco, a subsidiary of Eli Lilly, involved cows receiving 320mg, 640mg and 960 mg of BST at monthly intervals, beginning 13 weeks after calving. The highest response was obtained with the middle dose, producing a 21 per cent yield increase (*Animal Pharm* 1987).

The Institute for Grassland and Animal Production (IGAP) research farm at Shinfield near Reading is conducting a BST trial funded by Monsanto. The herd is permanently housed indoors and fed a mixed diet of concentrate and forage. The treated group is given fortnightly subcutaneous injections (over the shoulder using a 16 gauge needle) of 500mg prolonged release BST. Articles published as a supplement to the *Journal of Dairy Science* in 1988 describe the first year's results of the trial: milk yield was increased by 20 per cent and there was a 6 per cent increase in 'feed efficiency' (see, for example, Hard *et al.* 1988; Pell *et al.* 1988; Phipps 1987, 1988).

Another IGAP centre at Hurley near Maidenhead is in the third year of BST trials. This centre has contracts with Cyanamid as well as Monsanto. Results for the first lactation show a 20–28 per cent increase in milk yield, with a greater effect during early lactation when the Friesian herd was housed indoors and fed partly with concentrates. The cows reportedly show no signs of aversion to the injection, though there is a short-term swelling reaction at the injection site.

BST's less well-known property is as a growth promoter in calves and lambs. If BST is licensed, there will be no way to prevent farmers from using it to sidestep the EC hormone ban. The EC ban, which came into force on January 1 1988, prohibits the use of both natural and synthetic steroid hormones in meat production. BST is technically not covered by the EC directive because it is a protein hormone rather than steroid hormone.

The BST Working Party recommends that the MAFF should give consideration to the potential use of synthetic BST hormones as growth promoting agents. A MAFF pronouncement on this issue, in the context of the EC and UK steroid hormone bans, is needed.

How is BST Assessed?

MAFF's Veterinary Products Committee (VPC) was established under the terms of the Medicines Act (1968). This committee of academics and civil servants advises the Minister of Agriculture, Fisheries and Food on the licensing of new veterinary drugs. Monsanto, Cyanamid and Elanco are known to have applied to the VPC for commercial product licences for BST.

The assessment procedure has two stages. Firstly the product must be shown to be safe and of suitable quality, that is, chemically pure and stable. Companies must submit evidence that the product is safe in use for animals and humans. The nature of this evidence is, however, covered by official secrecy. The VPC then issues an Animal Test Certificate which permits larger scale trials. These trials are also subject to official secrecy. Their purpose is to collect data on the product's long-term effectiveness, and effects in general, before a commercial product licence is (or is not) granted. In the case of BST, VPC concern over possible animal health and welfare problems means that further research on these criteria is necessary (see below).

It is believed that in the UK some 1,000 cows are involved in two types of BST trials. Some trials are carried out by governmental institutions: Shinfield and Hurley Research Stations, and London University's Wye College in Kent, are known to have conducted BST trials. In these cases public resources are being used to further the interests of the drug companies, although this does mean that research results are more likely to be published.

The other form of BST trial uses commercial dairy farms. It is not known what financial arrangements drug companies make with the farmers involved in such trials. Farm trials are believed to have taken place in Devon, Somerset, Dorset, West Wales and Yorkshire, but no details of these have yet emerged. Whether these trials are run on an authentic commercial basis is unclear.

What Impact will BST have on Milk Composition?

There remain some unanswered questions about the composition of milk from cows treated with genetically engineered BST. The fat concentration of whole milk from BST-treated cows was substantially higher than milk from untreated cows in a number of studies (Bitman *et al.* 1984; Eppard *et al.* 1985a; Peel *et al.* 1982). This is most undesirable in view of current dietary advice to reduce intake of saturated fats.

Because the use of BST raises milk output considerably, there is concern that vitamin and mineral levels in milk from BST-treated cows may be lower than in milk from untreated cows (Brunner 1988b). According to Monsanto themselves, (Anon. 1988) prior to May 1988 only one paper had been published on milk minerals, and no data on vitamin levels had appeared in a scientific journal. Whilst BST use is restricted to trial production, these two potential problems do not affect the quality of milk in the general supply.

Results of tests on the residues of BST in milk from BST-treated cows have not been published in a scientific journal. (An invitation to the author to discuss drug company-sponsored research undertaken in a Belgian university was withdrawn.) Claims by the National Office of Animal Health (NOAH) – the veterinary drugs industry's pressure group – that BST levels in milk from treated cows are too low to measure, can only be viewed with scepticism.

Milk is regarded as a wholesome, unadulterated and nutritious product. The public must be confident that BST will affect neither milk composition nor flavour. The BST Working Party recommends that effective testing and monitoring is necessary to ensure that BST residues do not contaminate the milk supply or cow meat. The working party believes that BST milk residue test results should be available for public scrutiny. If residue testing is shown to be necessary, commercial approval for BST should not be countenanced until a routine test for BST residues in milk is available.

What Impact will BST have on Human Health?

Assurances have been given by NOAH that consumption of dairy products from BST-treated cows poses no risk to human health. NOAH advises that when BST is injected into humans it is inactive. Although this is so for the intact hormone molecule, it may not be so when the BST molecule is partially broken down. In one early study (Forsham *et al.* 1958), when a preparation of BST which had been exposed to a digestive enzyme was injected into a human subject, it showed human growth hormone (HGH) properties. This indicates that if it is present in milk in sufficient amounts, BST could possibly be active in the human intestine. A study has shown that injected synthetic HGH provoked immune reactions in 30 per cent of subjects (Crawford 1987). It is not known whether BST causes immune reactions in humans.

The level of at least one gut hormone – insulin-like growth factor (IGF) – which is active in humans is raised in the milk of cows (and goats) by BST. Levels of IGF in milk from cows given BST injections rose four-fold in one of the few studies so far reported (Prosser *et al.* 1988). This finding may have implications for babies who are given milk from BST-treated cows. IGF stimulates cell division, and is likely to exert its action on the walls of the intestine. According to Dr Colin Prosser, of the Institute of Animal Physiology and Genetics Research near Cambridge, IGF is present in breast milk and there is no evidence that its level in the milk from BST-treated cows is physiologically insignificant. He warns that 'the implications of IGF in milk for the human infant cannot be determined until we know more about the activity and function of milk IGF in the newborn' (Prosser 1988).

The BST Working Party recommends that full test results for each of the different BSTs and the dairy products from cows treated with them should be published before product licences are issued.

Are Consumers Drinking Milk from BST-Treated Herds?

Perhaps of greatest public concern is the fate of milk from BST-treated cows. Milk from BST-treated cows is added to the general milk supply, because, according to the MAFF, were it not safe to do so, BST would not have been granted an Animal Test Certificate. The practice of pooling milk from BST-treated herds for general consumption is not followed in any other European country. Neither the public nor the dairy industry know who is drinking the milk, for reasons of commercial confidentiality enshrined in the Medicines Act (1968). It can be argued that, in effect, the MAFF is protecting the drug companies from public scrutiny. The Retail Consortium, representing Co-operative Dairies, Express Dairies, Marks & Spencer, Sainsbury's, Tesco and Waitrose, wrote to the MAFF in September 1988 asking for the consumer's right to choose milk from cows not treated with BST to be respected. This would, at the very least, require separate farm collections and labelling of milk from BST-treated cows.

The BST Working Party recommends that the consumer's rights to know and to choose should be protected. Public confidence can only exist if secrecy is replaced by government openness. Urgent consideration should be given by the MAFF and the dairy trades to establishing mechanisms for accurate labelling of dairy products. The working party also recommends that the MAFF should produce the evidence for consumer safety of milk and dairy products from BST-treated cows and reveal the location of the trial farms.

What Impact will BST have on Employment?

The routine use of BST would be mischievous in a dairy industry plagued with over-production and milk quotas. BST's potential impact on farming in Europe has been only perfunctorily assessed. It is estimated that up to half the US dairy farmers may be bankrupted in ten years by the use of BST (Mix 1987).

The BST Working Party believes that an independent study of the impact of BST use in milk production is required to establish: effects on farming methods, including the use of grain and feed concentrates and indoor housing of dairy cows; effects on land use, notably the disuse of pastureland, and the potential effects on employment in the dairy trades.

What Impact will BST have on Animal Welfare?

There are fears that long-term use of BST may drive dairy cows beyond the limits of physiological endurance (Kronfeld 1987). The Farm Animal Welfare Council (FAWC), a MAFF committee, made their concerns public in December 1987 (Harrison 1987). A summary of research data up to June 1988 has been compiled by Compassion in World Farming (CIWF) (D'Silva 1988). Potential problems include lameness, diminished fertility, increased mastitis (udder infection) and other production-related diseases.

The BST Working Party believes that multi-lactation and multi-generation trials are needed to address these animal welfare concerns before BST is considered for commercial use. Potential BST-related disorders should be investigated independently. The concerns expressed in the FAWC report (Harrison 1987) on long-term animal welfare should be fully researched and allayed before commercial licences are issued for BST.

Will BST be Given a Product Licence?

At present in the UK all the evidence on the safety of veterinary pharmaceuticals is subject to official secrecy. The Medicines Act (1968) over-protects the drug companies, leaving the public to rely on the assurances of NOAH – the drug industry-funded pressure group – that BST is safe. NOAH has been misleading on key issues. For instance, it is inaccurate to say, as NOAH's publicity material claims, that the synthetic BSTs are identical to naturally occurring bovine growth hormone.

Just as importantly, the Medicines Act (1968), which permits only the criteria of safety, quality and efficacy to be used in the assessment of new products, may prevent the MAFF from refusing to license a product which many believe to be undesirable and unnecessary. Can we rely on the MAFF to defend the public interest? Former Minister of Agriculture, Fisheries and Food, John Selwyn Gummer commented: 'The idea that Britain should stand aside while allowing everybody else to produce milk in a modern way is barmy.' His mind was made up in mid-1988 even though the VPC did not make its recommendations to the MAFF until March 1989.

It is disturbing that a food-related product as sensitive as BST has been subject to so much misinformation and a lack of open debate. The BST Working Party believes that the issues of dairy economics, human health, animal welfare and public accountability must be resolved to the satisfaction of the public before a licence is granted for the commercial use of BST. The working party recommends a review of the Medicines Act (1968) to provide a sound regulatory framework for the licensing of novel products. The criteria which are currently used to evaluate a new product, namely safety, efficacy and quality, should be extended to include social and economic effects, and desirability.

Conclusions

Farmers are deeply divided on the issue of BST licensing. In the UK, at the Annual General Meeting (AGM) of the National Farmers' Union (NFU) in February 1989 a row broke out over BST policy. A vote resulted in 31 county branches supporting the call for a ban and 11 were undecided. At the AGM, NFU leaders were not swayed by delegates, believing that the BST trials should be completed before they decided on their position. However, a month later the

NFU Union Council responded to growing consumer concern over food production methods by voting for an EC-wide ban on BST.

In the European Parliament a worldwide ban on all growth hormones, including BST, was called for in September 1988. The Parliament's decisions are only advisory however, and the European Commission's position on BST has not been finalised. An expert committee of Commission officials, scientists and industry representatives are recommending 'controlled registration' of BST for use on approved dairy farms, which would be required to take advice from a veterinarian and a dairy nutritionist on administration, feeding and management regimes (*Animal Pharm* 1989).

If policy makers take the public's perception of BST seriously, then there are major problems ahead for those promoting this hormone. A report to the US National Dairy Board entitled *Consumer responses to the introduction of BST technology: devising a communications strategy* (1986), presents the results of six small group interviews with mothers. The report concludes that for a great many consumers the use of BST in milk production will seriously damage the image of milk as one of the very few remaining pure and natural foods, and that these consumers feel that they might reduce their consumption of milk significantly. The report asserts that the only way in which BST will be accepted by such consumers is if they can be persuaded that nothing has changed in their milk. However, the unanswered questions about the composition of milk from BST-treated cows and its long-term effects on human health militate against its acceptance by the public.

Other, wider, questions about BST also remain. How would BST affect employment in the industry, the health of dairy cows and the use of grazing land? BST is the first of a new wave of genetically engineered agricultural products. How would the granting of a product licence for BST affect the future of agriculture in a broader sense? There is a danger that powerful corporate interests will unduly influence the political decision-making process in the absence of informed public discussion of the issues. Public debate about the impact of genetic engineering on agriculture and food production is overdue.

9 The Welfare Implications of BST

Dr WOLFGANG GOLDHORN

Metabolism of the Dairy Cow

Bovine somatotropin (BST) is a genetically engineered growth hormone used as a drug to enhance the milk yield of dairy cows. The animal welfare implications of the use of BST become much clearer through an understanding of the way in which this sort of pharmaceutical enhancement of the metabolism of the cow alters her use of vital energy. Figure 9.1 illustrates the distribution of the total vital energy in cows under three different regimes, namely: in the wild, in intensive dairy farming, and in intensive dairy farming using BST. It compares the differing proportions of the total vital energy of the cow devoted to reproduction, self-maintenance and defence against disease, and the energy reserved for the 'emergency ration' for survival under critical conditions.

Undomesticated cows use most of their total vital energy for survival (see Figure 9.1(a)). Self-maintenance and defence energy is used to fight against adverse environmental conditions, hunger, parasites and other infectious agents, and against predators. Only a very small proportion of the total vital energy of undomesticated cows is used on reproduction because such cows usually produce only one calf per year and enough milk for the first few weeks of its life.

Over the centuries animal breeders have selected those cows that produce more milk than their calves require. As a result of this selective breeding the average annual milk yield per cow has increased from about 1,000 litres in 1900 to in excess of 4,000 litres today, of which an increase of 1,000 litres has occurred in the last 10–15 years. It is evident that today the major proportion of the total vital energy of high-yielding dairy cows is needed for milk production and only a small part is left for self-maintenance and defence, as illustrated in Figure 9.1(b). But what have been the welfare implications of this redistribution of the cow's vital energy?

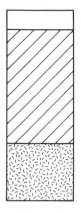 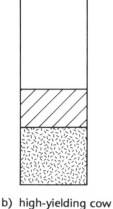 

a) undomesticated   b) high-yielding cow   c) high-yielding cow
   cow   in intensive dairy   in intensive dairy
                       farming   farming with BST

☐   Energy used for reproduction (calves and milk)

▨   Energy used for self-maintenance and defence against disease
    and adverse conditions

▨   Emergency ration of energy for survival under critical
    conditions

BST   bovine somatotropin

Figure 9.1:
**Uses of the vital energy of the undomesticated cow, the
high-yielding cow in intensive dairy farming and the high-yielding
cow in intensive dairy farming with BST-treatment**

Firstly, farmers have reconstructed their cowsheds in order to
create an environment more conducive to intensive dairy farming.
In order to sustain increased milk production they are obliged to
purchase concentrated commercially produced cowfeed instead of
feeding their herds on feed produced from their own crops, and
they have also had to intensify animal husbandry and management. The average life span of a cow has decreased because of an
enormous increase in metabolic disorders, fertility problems and
mastitis. Veterinarians have had to intensify their service to dairy
farmers through the use of new and more sophisticated drugs and
techniques which are increasingly expensive and carry new side-
effects.

Despite intensive dairy farming techniques, the proportion of the total vital energy of the cow required for survival in very critical situtations has remained intact – locked away as an emergency ration. This emergency ration remained inviolate, and there had been no access to it for any other purpose, not even milk production. With the advent of BST, the dairy industry now has the key to 'unlock' and use this emergency energy ration. This is illustrated in Figure 9.1(c).

Welfare of the Dairy Cow

Researchers do not know exactly what the effects of using BST on dairy herds will be because there has not been sufficient independent long-term research conducted so far. But one thing is absolutely certain: the considerable health and welfare problems associated with high-yielding dairy cows will be exacerbated. What sort of problems are likely to arise from the use of BST in intensive dairy farming?

First, the injections required at present to administer BST, whilst not very harmful, may create painful swellings and engender fear in the animal (Whitaker *et al.* 1988). BST-treated cows will not be permitted to graze for their food because high milk production can only be achieved with concentrated feed. Studies in West Germany indicate that only about 50 per cent of BST-treated cows respond with increased milk yield. It is likely that those cows which do respond to BST in this way will be selected for breeding, and nobody can predict what other qualities cows will lose by this single-criterion selection. With regard to disturbances of the dairy cow's metabolism, even without the use of BST to boost milk production high-yielding dairy cows already suffer from enormous health problems. With the use of BST there will be an increasing risk amongst dairy cows of metabolic disorders such as ketosis, acidosis and electrolyte imbalances. (Electrolyte imbalances are generally imbalances of sodium phosphate, potassium phosphate and calcium phosphate, which very often lead to infertility or severe, and sometimes even fatal, paralysis.) These metabolic disorders and electrolyte imbalances will lead to more fertility problems and, above all, will reduce dramatically the cow's natural resistance against infectious diseases.

Many viruses and bacteria inhabit the body of the dairy cow and her immediate environment. The change in the dairy cow's metabolic balance brought about by the administration of BST will tend to diminish her ability to resist infections such as infectious bovine rhinotracheitis (IBR), and those caused by certain species of Leptospira, Salmonella, Staphylococcus, Streptococcus, and

Trichophyton, some of which are quite dangerous to human beings, who are themselves becoming less resistant to microbial infections.

It is therefore likely that the administration of BST to dairy cows will diminish their welfare because they will suffer from an increase in metabolic disorders and microbial infections, and also from the side-effects of the treatments administered for these disorders and infections. Moreover, the use of BST will in turn increase the health risk to the consumer of dairy products as a result of the increased amounts of antibiotics and other drugs and prophylactics (preventive drugs and procedures) administered to BST-treated dairy cows.

Another, but indirect, implication of BST for animal welfare will be the fact that most smaller farmers will not be able to provide the 'high-tech' management required for BST administration. Family farms, where cows still have names and are almost regarded as members of the family, will be replaced by 'factory farms' with 'animal machines' for whom veterinarians will become their 'maintenance technicians'.

Role of the Veterinarian

The role of the veterinary profession is to act as a mediator between the interests of the farmers and those of the animals. It is from this mediating role that the profession derives its great credit and respect in society. Do veterinarians who betray their role as guardians of animals – those who remove the tails, lop the ears or cut the vocal cords of dogs, declaw cats, make laparotomies in pigs for embryo transfer, or inject hormones into healthy farm animals to provoke hyperovulation or higher milk yield – still merit this respect? When traditional family farming turns into agribusiness, will veterinarians go along with this and become reckless business people or will we be mindful of our cautionary function and professional responsibility? The veterinary profession as a whole must reconsider its role in the light of such questions as: is feasibility adequate grounds for doing something? If something is legal does this mean it is also moral? Is it sufficient for veterinarians to protect animals from pain, harm and suffering? Should they not safeguard the identity and dignity of animals as well?

The above questions, however, do not only concern farmers and veterinarians but everybody, because everything we do in the world has repercussions on ourselves and will change our self-perception as human beings. Society cannot leave these fundamental questions to scientists or politicians or certain professions.

In 1965 the Brambell Report was published (Brambell 1965). This report, which was commissioned by the British Government in response to animal welfare concerns over the use of intensive methods of keeping livestock – popularly known as 'factory farming' – became, and still is, an important report which is used by animal welfarists all over Europe. There is a need for a new Brambell Report on the dignity of animals, because even in breeding with traditional methods we have gone too far, especially in the breeding of pet animals.

Farmers and veterinarians must be brought together with philosophers, theologians, consumers and those interested in welfare issues to form what Michael Fox has called 'Bio-ethics Commissions'. Such commissions would be permanent and would address matters arising from these questions, in the hope that a consensus of opinion might be formed which would indicate to potential investors whether or not a certain product or technique would be acceptable to society. If the consensus view on a particular research development was that it would not be acceptable to the public then this would be a signal to investors not to invest money in this kind of research. Asking for legislation to control such technological developments when a product is already on the market is too late. Taking BST as an example: if the companies which have developed BST had known that farmers, veterinarians and consumer associations would be opposed to its use would they have invested so many millions of dollars in it?

BST, a growth hormone produced in genetically engineered bacteria, is only a pilot project in the diffusion of the new genetic engineering technology into farming. The new techniques of genetic engineering are being used to manipulate growth through the use of growth hormones even though the process of growth itself is not fully understood. Researchers are working on improving the efficiency of farm animals by manipulating the genetic code of the animals themselves. Such genetic engineering developments in farming will have even greater and more unpredictable impacts on animal and consumer welfare than BST.

Cornucopia or Pandora's Box? Any new technological development in our world always contains the potential for both. Progress should not be prohibited, but defined and controlled. We could not have prevented Einstein from finding his famous equation, $E=mc^2$, but we can learn from the many mistakes made in defining the limits and conditions of its practical application. We should do so *before* the first biogenetic Chernobyl occurs. The solution to regulating new technology is not to prohibit research but to facilitate regulatory agencies and researchers by making available to them the critical guidance of dialogue with a permanent public body, such as a Bio-ethics Commission.

10 Why BST Must be Opposed

Dr MICHAEL FOX

To oppose the use of bovine somatotropin (BST) is neither an anti-science nor an anti-progress neo-Luddite reaction. There are many valid reasons for opposing the use of this new product of biotechnology in dairy cattle. These reasons fall into three broad categories: social and economic issues; bio-ethical concerns, and consumer health risks.

Social and Economic Issues

Firstly, however, we might ask who needs BST? It would seem that the primary, if not sole, beneficiaries would be its manufacturers and retailers. Benefit to consumers would be insignificant since milk prices would be kept at a stable level – this is the situation now in the USA and the European Community (EC) where huge milk surpluses are bought and stored by governments at public expense to maintain price supports for the dairy industry.

The use of BST by the dairy industry will not mean that consumers will benefit from an ever cheaper and more plentiful supply of milk. Production costs per cow will increase and not decline because, as I describe below, there are hidden costs in the use of BST that must be accounted for. The price of milk and other dairy products is not likely to decrease for consumers, who today pay the dairy industry billions of dollars in the form of government price supports, and for the purchase and storage of surplus dairy products. If no more cows were milked today in the USA, it has been estimated that there is enough stored produce to supply every person in the USA with powdered milk for two years and cheese for four months. BST is a product of biotechnology that society should reject, since the only beneficiaries will be its manufacturers and the larger dairy farms of today that, with the advantages of economies of scale, will become the monopolists of the dairy industry tomorrow. It is on these 'factory farms' that the dairy cow will suffer in the name of 'progress'.

If BST is allowed onto the market, and is not prohibited by dairy farmer associations, a new competitive economic treadmill will be started. Once one farmer starts to use BST and gains a slight economic edge over competitors, others will follow suit. Some analysts predict that this will lead to the demise of smaller scale family farm dairy operations thus disrupting the structural stability of the dairy industry. The preservation of family farm dairy operations is more than just an aesthetic and a cultural concern. Preserving the diversity and flexibility of farming practices is part of maintaining a structurally sound and ecologically stable agriculture.

Large monoculture 'factory farms' and farming practices, protected and encouraged by government subsidies, commodity price supports and inappropriate biotechnological innovations, will be the nemesis of modern agriculture. Non-therapeutic animal pharmaceuticals such as BST, designed to make farm animals ever more 'efficient' and 'productive', are touted by their proponents as safe and naturally occurring proteins, peptides and biologics. The potential for causing harm to the animals themselves is equalled only by the probability of harm to the structure and future of agriculture itself. It is ludicrous for bureaucrats and biotechnocrats to claim that, with government regulations and oversight, all will be well, that the welfare of animals would be assured, and that the future of a sustainable, ecologically sound agriculture would not be jeopardised. Such a claim reflects, at best, a condition of historical amnesia with respect to the past use and abuse of antibiotics, hormones and anabolic steroids in farm animals, and of herbicides and pesticides on crop plants. Furthermore, such regulatory oversight of the use of BST would be an additional hidden cost in the real price of milk and other dairy products, and would be borne by the consumer.

Aside from these socio-economic concerns, which are being hotly debated by dairy industry and agricultural analysts (Kalter *et al.* 1984; Kalter 1985; Browne 1987; USDA 1987), the impact of BST on the health and welfare of dairy cows has received scant attention, and adverse consequences have been virtually concealed in industry-supported reports by animal production scientists. Treatment with this hormone will mean that dairy cows will be under even greater production pressure. Their body metabolism, which is already being pushed to the limits on 'factory farms', will be stressed further. Current intensive dairy husbandry practices are responsible for a host of so-called production-related diseases (see Fox 1983; Harvey 1983), which result in stress and suffering to dairy cows that, as a consequence, 'burn out' in three to four years.

BST will be used in conjunction with these other intensive dairy-factory practices. The cow will become a production-machine, spending most of her life eating almost constantly in order to maintain the quantity of milk production that selective breeding, energy-rich concentrate feeding, and BST-treatment evoke.

BST-treatment, as documented by veterinarian David S. Kronfeld (1988), will intensify the cruelty to which 'factory farmed' dairy cows are subjected today. Their physiology will be under even greater stress and this will result in suffering when their systems break down and they succumb to such production- and husbandry-related diseases as mastitis (painful infected udders), crippling lameness, fatty liver disease, and metabolic disorders such as ketosis and acetonemia. Bitman *et al.* (1984) reported a 41 per cent increase in milk fat secretion following BST-treatment, the milk fat being derived from the animals' own adipose (fatty) tissue reserves. A negative energy balance thus results and since the mineral elements in milk from BST-treated cows are maintained within the normal range (Eppard *et al.* 1985b), the drain on the cow's reserves will be significant. Her resistance to infectious diseases will also be lowered, which will make her more susceptible to sickness and increase the probability of her suffering.

Only a few studies on a small number of dairy cows have been conducted on the animal welfare implications of BST-treatment, and even these studies have not been carried out for any significant period of their lives. Kronfeld (1988) notes 'favorable responses to BST have been presented promptly, loudly and repeatedly. Unfavorable results have been delayed, subdued and obscured.' He has documented studies showing increased health problems in BST-treated cows, ranging from mastitis and reduced disease resistance, to reduced fertility. He suggests that farmers (one-quarter of whom are predicted to be pushed out of business during the first three years of BST use) will require additional management skills to deal with these new problems and risks. The need for additional management skills increases the probability of human error. 'Agribiotechnologies' should be designed to minimise such dependence upon managerial skills. If they are not, the one at greatest risk will be the dairy cow herself.

As veterinarian Rolf Kamphauser (1988) has emphasised, the complex network of hormonal and metabolic systems and cycles of dairy cows are little understood, whether under normal, healthy conditions or under those of the 'factory farm'. Against this background of a paucity of basic knowledge, claims about the safety and efficacy of BST in dairy cows on an individual and long-term, even generational, basis, as well as under different husbandry conditions, are at best unreliable. For example, in cows selectively bred for very high milk production the use of BST may not increase milk yield significantly. Moreover, since BST is not the only hormone that regulates milk production, and since the neuroendocrine system (nerves and hormones) is a complex network, it is naive to expect that BST will not affect systems and processes other than lactation, notably the reproductive and immune systems. For instance, an insulin-resistant state has been observed in dairy cows following growth hormone treatment. Moreover cows undergoing BST-treatment exhibit poor heat tolerance, which makes the use of BST in tropical conditions counterproductive.

The potential for the misuse of BST, and of consequent animal abuse, is very real. In my opinion the most obvious animal abuse, which the 'real world' economics of the competitive dairy industry would virtually guarantee, is this: dairy cows would be hyperstimulated with BST when they are in poor condition or when their lactation cycle begins to wane. The tragedy for the dairy cow is that she naturally expends her body resources and reserves to produce milk in order to ensure the survival of her calf. Her health and welfare will be jeopardised when BST is used at her expense to ensure the survival, not of her calf, but of the dairy farmer caught on the competitive economic treadmill of a BST-addicted dairy industry. This would be done regardless of her physical condition, which would be allowed to deteriorate, because she is destined to be slaughtered and turned into pet food and leather goods anyway. However, it is unlikely that any research study would model this real-life, dairy 'factory farm' potential misuse of BST.

Some proponents of BST claim that BST-treated cows need no extra food. However, this claim is only a half-truth. Increased milk production without an increase in food intake will result in a negative energy balance for the cow [see also Chapter 9]. This will lead to the cow's increased susceptibility to production-related diseases. Kalter *et al.* (1984) have emphasised how the use of BST will change the kind of feed-crops grown for dairy cows, since these animals will require more high-energy concentrates in their diet. In summary, the claim that BST-treated cows need no extra food is incorrect, and underscores the erroneous belief that you can get something for nothing.

BST is a product that the advanced industrial nations do not need. It will not make dairy cows significantly more efficient since they will need more feed to make more milk. But it will mean that fewer cows will be required to satisfy the nation's need for milk and dairy products, hence its appeal to those who are imposing the value of industrial efficiency on the dairy industry regardless of its adverse consequences to the farmer and the cow. However, since dairy cows treated with BST will have shorter productive lives, profits will be offset by the need to replace these cows at a faster rate and it is costly to raise replacement dairy calves. *This hidden cost has not been considered by proponents of BST.*

Consumer Health Risks

The treatment of cows with BST is also controversial because of the potential consumer health risks arising from the consumption of milk and dairy products from these cows. It is claimed that milk from BST-treated cows differs in content from milk from non-BST-treated cows. According to Torkelson *et al.* (1987) milk from BST-treated cows may contain BST. Such milk is therefore adulterated. Nothing is known about the potentially harmful effects of break-down products of BST on people and developing foetuses. It has also been observed that the butterfat obtained from the milk of BST-treated cows contains a higher proportion of long-chain fatty acids and a lower proportion of medium- and short-chain fatty acids than butterfat from untreated cows (Eppard *et al.* 1985b). An important consumer concern about the consumption of milk from BST-treated cows arises from the adverse effects of long-chain fatty acid consumption on the health of certain individuals (see, for example, Moser, H.W. *et al.* 1981; Moser, A.B. *et al.* 1987). For example, dairy products from BST-treated cows could be harmful to persons already suffering from a variety of malabsorptive disorders in which long-chain fats are poorly digested and absorbed (Hashim 1967).

Conclusions: The 'New Creationism'

Today's technocratic attitude toward biological systems in general and the dairy cow in particular needs to be questioned and challenged. An industrial, mechanistic paradigm cannot be safely imposed upon biological systems. In the absence of respect for, and understanding of, the nature and integrity of biological systems they become dysfunctional and diseased, necessitating new bio-technological, legislative and other palliatives which are incorrectly regarded as progressive innovations and refinements. An editorial in the *Christian Science Monitor* (1988) entitled 'Beware the "New Creationism"' gave the following warning:

> As scientists, businessmen, and the public look at the promise of genetic engineering, they need to constantly guard against a subtle and dangerous attitude. It could be called the new creationism.
>
> Its impact on human advancement can be just as detrimental as the 'old creationism'. This holds that a Supreme Being was responsible for creating the material universe and a hierarchy of

distinct forms of life in a very brief span of time several thousand years ago.

The new creationism replaces the Deity with human beings wielding the tools of molecular biology. Though focusing solely on organic life, the new creationism assumes responsibility for 'creating' improved plants and animals strictly for human benefit in a fairly short time. In essence, new creationism threatens to erode humanity's respect for the intrinsic individuality and value of all forms of life. As subtle forms of new creationism take hold on thought, they can lead to the kind of hubris reflected in the summary of a recent genetic engineering article: 'Regulators and the public are inhibiting a major new industry. They must better understand the potentials and limits of biotechnology, and then get out of the way.'

Get out of the way, indeed – as if those outside a tight circle of experts have nothing of value to say or no concerns to raise. Hubris is perhaps the most blatant sign that new creationism is gaining a foothold. More subtle is this attitude's tendency to foster a mechanistic view of animals and humans. Avoiding such an outcome demands constant vigilance and constant questioning on the part of scientists, businesspersons, and the broader public alike.

The rising technocracy, with its materialistic values of economic determinism, scientific imperialism and secular humanism, is turning the natural world into an industrialised wasteland and the cow into a biomachine. Public opposition to BST represents a growing and healthy opposition to this trend in human consciousness. The fundamental question, therefore, is: are we to stand by and see the natural world, the cow and other creatures re-made into some profane image of industrial efficiency and productivity, or are we to become creative, conscious participants in the natural expression and fulfillment of life on this planet in all its beauty, harmony and diversity? As Thomas Berry (1988) has eloquently argued, the choice is ours and it is our collective will that determines the future of humanity and the fate of the earth. The cow, long revered by pre-industrial civilisations, is now in as much jeopardy under our dominion as the earth itself. To oppose BST, therefore, is more than a sentimental or humanitarian gesture. It is enlightened self-interest.

Discussion II

Joyce D'Silva, the Athene Trust: I would like to point out that all the references in the Compassion in World Farming (CIWF) *Dossier on BST* [D'Silva 1988] are from acknowledged veterinary and scientific sources and I therefore consider that Mr Deakin is not justified in calling it a 'misinformation dossier'. In response to Mr Deakin's statement that BST is species-specific, I understand from scientific experts that BST is not species-specific but species-limited, which is rather different.

Eric Brunner, LFC: Mr Deakin has accused the London Food Commission (LFC) of 'gastronomic terrorism' because its BST Working Party Report [Brunner 1988a] cites a scientific reference which refers to raised fat levels in milk from BST-treated cows in a short-term trial. However, if he is proposing that BST be used as a 'management tool', then presumably BST *will* be used for the short-term treatment of cows.

In the LFC report on BST, we have tried to be fair when looking at the literature on BST. Allow me to read just one paragraph from the report.

> The National Office of Animal Health (NOAH) claims that milk composition is not changed by BST use. This claim is supported by some studies (Bines and Hart 1982; Faras and Muscoplat 1985; Bauman *et al.* 1985; Phipps 1987), but not by all of those published. A study of the milk fat profile by Bitman *et al.* (1984) of Holstein cows given daily BST injections (51.5 International Units subcutaneously) in a 28-day crossover trial showed a 27 per cent higher fat concentration in the milk of BST-treated cows compared to controls [Brunner 1988a].

There are two other papers at least which support the view that fat levels in milk from BST-treated cows are raised. This occurs in short-term trials in cows which have recently been injected with BST [Eppard *et al.* 1985a; Peel *et al.* 1982]. Finally, I think that hys-

93

terical over-reactions like that of Mr Deakin's in response to the LFC's report on BST do no good for the case of the drug companies at all.

Irene Williams, CIWF: Mr Deakin, you said that BST had no detrimental effect on the welfare of cows. But you omitted to tell us how many injections have to be given to these cows and where on the body they have to be given. You also did not tell us that cows have to be isolated and are not allowed to graze.

May I say also that we in CIWF are not impressed by the fact that two Townswomen's Guild ladies did not know the difference between milk from BST-treated cows and other milk. It is our job to tell these people the difference between BST-derived milk and the rest, and we shall do it very effectively.

Robert Deakin, Monsanto: Each cow receives an injection a maximum of once every two weeks. There are a number of efforts to produce a more sustained or longer acting BST product. The BST syringe is about the size of a pencil, which represents the volume of BST which is injected into the cow. The inoculum is a mixture of BST and sesame seed oil and looks like mayonnaise.

There are a number of studies which address the question of grazing. One in particular was undertaken under the auspices of the Agriculture Department Advisory Service (ADAS) – an extension service of the Ministry of Agriculture, Fish and Food (MAFF) – and published in the UK [Furniss *et al.* 1988]. In this trial, which was conducted during the winter housing period, both BST-treated cows and untreated cows received the same flat rate of feed concentrate. The treated animals responded to BST and did not suffer in any way in terms of body condition and score weight. In the spring the cows were turned out to pasture. All of their milk levels declined, of course, but the difference was maintained and the BST-treated cows continued to put on body weight. BST-treated cows were perfectly capable of producing milk when fed on pasture. Studies have also been done in Australia with cows on pure pasture who responded to BST-treatment without any adverse effects on their welfare [Peel *et al.* 1985].

Ruth McNally, Bio-Information (International) Limited: The UK BST trials are being conducted under Animal Test Certificates issued by the Ministry of Agriculture, Fisheries and Food (MAFF) under the Medicines Act (1968). The purpose of the trials is for the pharmaceutical companies to gather data on the efficacy of administering genetically engineered BST to dairy cows. Why is it that in the course of gathering this data, milk from cows in BST-treated herds is being distributed to consumers who are not in a position to discriminate against it?

Robert Deakin, Monsanto: The procedure for obtaining an Animal Test Certificate for BST trials requires that first of all you demonstrate whether or not there will by any residues in the milk and whether or not those residues could have any effect on the consumer of the milk issuing from the animal under test. If there is a potential for residues which is as yet undetermined, then the milk cannot be distributed. It was demonstrated conclusively to the authorities in the UK and to the Food and Drug Administration (FDA) in the USA and to others that the milk from BST-treated cows was unchanged, and even if it were not unchanged, there would be no effect on the health of humans consuming dairy products from BST-treated cows because BST is a protein which is not active in humans. And so the authorities decided that, in the case of BST, milk safety is not an issue. What is an issue for these BST trials is the long-term effect on dairy cows of BST-treatment, whether it is efficacious, whether it is a worthwhile product, and whether it does the job that it is supposed to do.

Ruth McNally, Bio-Information (International) Limited: If the companies testing BST are confident that the consumer will not discriminate between 'normal' milk and milk from cows in the BST trials, I am puzzled as to why that milk is being distributed and sold without being labelled as coming from BST-treated cows.

Robert Deakin, Monsanto: The milk from BST-treated cows is not labelled because it is not different. Milk is milk is milk, and the milk from the treated cows is not different. In the UK milk goes to the Milk Marketing Board (MMB) where it is pooled; people get the impression that somehow milk from BST-treated animals is being spread all over the country. But the bottom line is that the milk from BST-treated cows is not different and therefore you cannot label something that is not different.

Ruth McNally, Bio-Information (International) Limited: Whether or not the milk from BST-treated cows is or is not different from normal milk, and whether or not it should be labelled as such is in dispute. I believe the sale of unlabelled milk from BST-treated cows with the consent of the MMB to be a clear case of 'producer sovereignty' [where producers impose their products on the consumer market, and consumers have no way of discriminating against them].

Dr Caroline Murphy, RSPCA: I refer to the sale of milk from cows that have been treated with BST. It worries me very greatly that if Monsanto's attitude is that the milk from such cows is in no way different from ordinary milk this will mean that Monsanto is not actually doing any follow-up studies on consumers in order to show that this is the case. One of the consequences of the British system of milk distribution from BST-treated cows into the main milk supply is that UK consumers are exposed to whatever consequences there are and have no information on what those consequences might be. Could you tell me whether Monsanto is doing follow-up studies on consumers of milk from BST-treated cows?

Robert Deakin, Monsanto: The answer is yes. We are doing more studies. Extensive rearch is being done on the milk itself and the chemical analysis of the milk.

Dr Caroline Murphy, RSPCA: Is Monsanto doing research on the long-term consequences of consuming large quantities of milk from BST-treated cows, for example, by pregnant women and newborn babies?

Robert Deakin, Monsanto: We do not do studies on human beings and we have not injected any people with BST. What we are doing are studies with rats which are the typical model system. We establish the amount of milk that a baby would consume at birth – let's say a typical baby would consume a litre of milk in a day. Then we determine how much of the various constituents of milk, for example, BST, would be contained in that litre of milk. We then test those constituents on rats at dose levels of 200–2,000 times the levels expected in a litre of milk, to determine whether the constituents are absorbed or have any harmful effects.

Dr Caroline Murphy, RSPCA: I must say I find it morally reprehensible that Monsanto is not doing long-term studies on the consequences to people of drinking BST-treated milk, and I am quite astonished that you are not even planning to do them.

Dr Neil Craven, Monsanto: Mr Brunner raised a number of concerns about the quality and safety of milk from BST-treated cows to which I would like to give some answers which are publicly available in the literature. I think Mr Brunner would agree that in reviewing the scientific literature you have to consider all the reports to get a balanced view.

Insulin-like growth factors (IGF) in milk were mentioned. At a conference in Cambridge two weeks ago [September 1988], Mr Brunner was publicly given specific detailed answers on this point. The IGF levels in milk after BST-treatment remain substantially lower than those found naturally in cow milk in early lactation, and lower than levels found in human breast milk. Indeed, the net biological activity of such factors in milk varies considerably according to the age of the cow. Should we be concerned whether we drink the milk from a four-year old cow rather than a three-year old cow? I think not, but the age of the cow has a greater influence on IGF activities than BST-treatment does.

Mr Brunner mentioned that in the literature he had come across a report from 1958 [Forsham *et al.* 1958] which implied that partially digested BST retained significant activity in humans. Since this work was done there have been many attempts to repeat this result, all of which have proved unsuccessful, and I have with me 20 references to published work in scientific journals published between 1968 and 1986 which testify to this. Is Mr Brunner aware of these 20 references which show that BST does not retain significant activity after partial digestion, and if not, why not?

Eric Brunner, LFC: Dr Craven, from Monsanto, has asked me why I was not aware of these 20 references in the BST literature. The reason is because I work for the LFC and not for the MAFF, or the European Commission, or the British Industrial Biological Research Association, or the drug companies which are interested in promoting BST. My role is to take the evidence that comes before me. Evidence came before me at the Cambridge Conference referred to, which took place two weeks ago, and now that I have been told that there have been 20 studies that have not been able to reproduce the result of the 1958 study on BST activity after partial digestion, I would like to see reports of those 20 studies. On the other hand, at the same conference I talked to Dr Craven and asked him to send me references to these studies and the data on vitamin and mineral composition of milk from BST-treated cows, but he did not do so.

Ken Nobus, dairy farmer, USA: My brother and I run a family farm with a herd of 200 dairy cows. We have been testing BST for Monsanto on our dairy herd over the past ten months. I would like to ask Dr Fox and Dr Goldhorn where have they seen these cows that have been so drastically altered, and gone down in physical condition due to the administration of BST?

In our herd we have not seen any of the results you are talking about. We have never seen a case of ketosis, we have not seen an increase in the number of lame cows, we have not seen an increase in mastitis, nor indeed have we seen any ill-effects at all. On the other hand, we have seen some beneficial effects. We believe the cows are more healthy if they are on BST. We are not running them into the ground, indeed we think that we could manage them even more intensively than we do and get more yield without harming them.

Dr Michael Fox, HSUS: Your Monsanto BST trial is only in its tenth month and we need data on two or three lactation cycles. I would also like to see comparisons between your farm and other farms. I like your herd size and the fact that your farm is a family farm. My fear is that we are going to see some of the real abuses of BST when it is used by the larger 'dairy factories' where there are 2,000–5,000 dairy cows.

I am concerned about you good people. I am not saying you are good because you use BST but you, in my category, are an endangered species. My big concern is that 'dairy factories' are going to take over the family dairy farms and they are going to 'burn the animals out', and you are here supporting the agrichemical industry that will probably kick you in the ass six months down the line.

Al-Hafiz B.A. Masri: I am more interested in these problems from the moral and ethical points of view. I wish to ask Mr Deakin whether those scientists who are recommending BST have taken into consideration the ethical point of view, namely, that even a cow has the birthright to enjoy her meals in the open air as nature intended her to do.

Robert Deakin, Monsanto: That is a very romantic idea with which, however, BST is entirely consistent, particularly in the EC where dairy farmers are constrained by their milk quota. Were such farmers to use BST, in some cases they would be able to reduce the number of cows in their herd. They would thus be able to provide more space for and give more attention to the cows that are left in the herd.

I think the days are gone when a pastoral sort of life for cows was possible. Cows and milk have to be produced in a commercial setting to feed the population of the world. The cows have to be subjected to a high level of animal husbandry. The farmers have to take good care of their cows because they are their prime investment. It should be done compassionately, and with BST I think it can be.

Jon Wynne-Tyson: I am making a plea rather than asking a question. My primary concern is with the suffering we inflict on other forms of life. My concern is not only on behalf of the animals, but for the human beings whose humanity is diminished by their treatment of other life. I would like to suggest that the debate should not be restricted to whether BST and other biomedical products are safe and practicable, or whether genetic work on animals must be controlled to ensure their welfare. Where scientists work on other living creatures in pursuit of power and profits, their own or their employers', it will prove impossible, then as now, to ensure the welfare of the animals concerned. Only ceasing to exploit animals can ensure that they no longer suffer.

Anon: We should not tolerate the fact that in the UK milk from BST-treated cows is being pooled with other milk, and consumers are being treated as human guinea pigs. I would like to propose a resolution that this conference demands that milk from BST-treated cows be labelled as such until this milk is proven to be no different from ordinary milk [see 'Conference Resolutions'].

References II

Animal Pharm (1987), 20 November, no. 142 (Surrey: PJB Publications).

Animal Pharm (1989), 17 February, no. 173 (Surrey: PJB Publications).

Anon. (1988) *Quality and Manufacturing Properties of Milk from BST-Supplemented Cows* (Brussels: Monsanto Animal Sciences Division).

Asimov, G.J. and N.Z. Krouze (1937) 'The lactogenic preparations from the anterior pituitary and the increase of milk yield in cows', *Journal of Dairy Science*, vol. 20, p. 289.

Barbano, D.M. *et al.* (1988) 'Influence of methionyl somatotropin on general milk composition', *Journal of Dairy Science*, vol. 71 (suppl. 1), p. 101.

Bauman, D.E. *et al.* (1985) 'Responses of high-producing dairy cows to long-term treatment with pituitary somatotropin and recombinant somatotropin', *Journal of Dairy Science*, vol. 68, pp. 1,352–62.

Bauman, D.E. (1987) 'Bovine somatotropin: The Cornell Experience', *Proceedings of the National Invitational Workshop on Bovine Somatotropin*, pp. 46–56, USDA Extension Workshop.

Bauman, D.E. *et al.* (1989) 'Long-term evaluation of a prolonged-release formulation of N-methionyl bovine somatotropin in lactating dairy cows', *Journal of Dairy Science*, vol. 72 (forthcoming).

van den Berg, G. and de Jong, E. (1987) 'The influence of the treatment of lactating cows with methionyl bovine somatotropin on milk properties', Nederlands Instituut voor Zuivelonderzoek, December, 1986.

Berry, T. (1988) *The Dream of the Earth* (San Francisco: Sierra Books).

Bines, J.A. and Hart, I.C. (1982) 'Metabolic limits to milk production, especially roles of growth hormone and insulin', *Journal of Dairy Science*, vol. 65, pp. 1,375–89.

Bitman, J. *et al.* (1984) 'Blood and milk lipid responses induced by growth hormone administration in lactating cows', *Journal of Dairy Science*, vol. 67, pp. 2,873-80.

Boyle, C. and Wheale, P.R. (1986) 'Philosophy and sociology of science', in C. Boyle *et al.*, *People, Science and Technology* (Hemel Hempstead: Wheatsheaf; New York: Barnes & Noble) pp. 26–43.

Brambell, F.W.R. (1965), *The Brambell Report*, (London: HMSO, Cmnd 2836) reprinted 1967.

Browne, W.P. (1987) 'Bovine growth hormone and the politics of uncertainty: fear and loathing in a transitional agriculture', *Agriculture and Human Values*, Winter, pp. 75–80.

Brumby, P.J. and Hancock, J. (1955) 'The galactopoetic role of growth hormone in dairy cattle', *New Zealand Journal of Science and Technology*, vol. 36A, p. 417.

Brundtland, G.H. (1987) *Our Common Future* (United Nations World Commission on Environment and Development).

Brunner, E.J. (1988a) *Bovine Somatotropin: A Product in Search of a Market* (London: London Food Commission).

Brunner, E.J. (1988b) 'Safety of bovine somatotropin', (letter), *Lancet*, vol. 2, p. 629.

Buckwell, A. and Morgan, N. (1987) *Impact of Bovine Somatotropin in The United Kingdom in The Context of Milk Quota* (Department of Agricultural Economics, Wye College, University of London).

Chalupa, W. and Galligan, D.T. (1989) 'Nutritional implications of somatotropin for lactating cows', *Journal of Dairy Science*, vol. 72 (forthcoming).

Crawford, M. (1987) 'Genentech sues FDA on growth hormone', *Science*, vol. 235, pp. 1,454–5.

Christian Science Monitor (1988) 'Beware the "New Creationism"' editorial, 3rd June.

de Hoop, D.W. *et al.* (1987) *Economische effecten van het gebruik van bovine somatotropine in Nederland* (Landbouw-economisch Instituut, Afdeling Landbouw, 's-Gravenhage).

Dickman, S. (1989a) 'Europe delays BST decision', *Nature*, vol. 340, p. 415.

Dickman, S. (1989b) 'EC announces bovine hormone moratorium', *Nature*, vol. 341, p. 274.

D'Silva, J. (1988) *Dossier on BST* (Hampshire: Compassion in World Farming).

Eppard, P.J. *et al.* (1985a), 'Effect of dose of bovine growth hormone on lactation of dairy cows', *Journal of Dairy Science*, vol. 68, pp. 1,109–15.

Eppard, P.J. *et al.* (1985b) 'Effect of dose of bovine growth hormone on milk composition: Alpha-lactalbumin, fatty acids, and mineral elements', *Journal of Dairy Science*, vol. 68, pp. 3,047–54.

Eppard, P.J. *et al.* (1987) 'Effect of 188-day treatment with somatotropin on health and reproductive performance of lactating cows', *Journal of Dairy Science*, vol. 70, p. 582.

European Commission (1988) 'The use of somatotropins in animal production in the European Community', summary of Community Seminar on *Somatotropin in Livestock Production* (Brussels: Commission of the European Communities, Directorate General for Agriculture - VI/8/11.2).

Faras, A.J. and Muscoplat, C.C. (1985) 'The impact of genetic engineering on animal health and production', *Journal of Animal Science*, vol. 61, suppl. 2, pp. 144–53.

Forsham, J. *et al.* (1958) 'Nitrogen retention in man produced by chymotrypsin digests of bovine somatotropin', *Metabolism*, vol. 7, pp. 762–4.

Fox, M.W. (1983) *Farm Animals: Husbandry, Behaviour and Veterinary Practice* (Baltimore: University Park Press) pp. 111–12.

Furniss, S.J. *et al.* (1988) 'Milk production, feed intakes and weight change of autumn calving, flat rate fed dairy cows given two-weekly injections of recombinantly derived bovine somatotropin (BST), *Animal Production*, vol. 46, p. 483 (abstr.).

van der Giessen, L.B. (1986) 'Structurele ontwikkelingen in de veehouderij: een blik in de toekomst', *Bedrijfsontwikkeling jaargang*, vol. 17–19, pp. 270–3.

Hard, D.L. *et al.* (1988) 'Effect of long term sometribove, USAN (recombinant methionyl bovine somatrotopin) treatment in a prolonged release system on milk yield, animal health and reproductive performance-pooled across four sites', *Journal of Dairy Science*, vol. 71 (suppl. 1), p. 210 (Abstr.).

Harrison, R. (1987) 'Letter to Mr Michael Jopling', *Official Report (Hansard)*, 17 December, cols. 740-1.

Hartnell, G.F. (1987) 'Evaluation of vitamins in milk produced from cows treated with placebo and methionyl somatotropin in a prolonged release system', *Monsanto Technical Report* MSL 5429.

Harvey, G. (1983) 'Poor cow', *New Scientist*, 29 September, pp. 940–3.

Hashim, S.A. (1967) 'Medium-chain triglycerides – clinical and metabolic aspects', *Journal of the American Dieticians Association*, vol. 51, p. 221.

Johnsson, I.D. *et al.*(1987) 'The effects of dose and method of administration of biosynthetic bovine somatotropin on live-weight gain, carcass composition and wool growth in young lambs', *Animal Production*, vol. 44, pp. 405–14.

Kalter, R.J. (1985) 'The new biotech agriculture: unforeseen economic consequences', *Issues in Science and Technology*, Fall, pp. 125–33.

Kalter, R.J. *et al.* (1984) *Biotechnology and the Dairy Industry: Production Costs and Commercial Potential of the Bovine Growth Hormone* (Ithaca, New York: Cornell University Department of Agricultural Economics) A.E. Research 84-22.

Kamphauser, R. (1988) *BST and its Effects on Health* (Bonn, West Germany: The Enquete Commission Estimation and Evaluation of the Impact of Technology).

Kimbel, S. *et al.* (1987) *Bovine Somatotropin (BST)* (Alexandria, Va.: Animal Health Institute), June.

Kirchgebner, M. *et al.* (1988) 'Effect of bovine growth hormone on energy metabolism of lactating cows in long-term administration', in *11th Symposium on Energy Metabolism*, EAAP Publication, 18–24 September.

Koldovsky, O. and Thornburg, W. (1987) 'Hormones in milk', *Journal of Pediatric Gastroenterology and Nutrition*, vol. 6, pp. 172–196.

Kronfeld, D.S. (1965) 'Growth hormone-induced ketosis in the cow', *Journal of Dairy Science*, vol. 48, p. 342.

Kronfeld, D.S. (1987) 'The challenge of BST', *Large Animal Veterinarian*, November/December, pp. 14–17.

Kronfeld, D.S. (1988) 'Biologic and economic risks associated with use of bovine somatotropins', *Journal of the American Veterinary Medical Association*, vol. 192, no. 12, 15 June, pp. 1,693–6.

Langbehn, C. and Wahlers, H.W. (1987) *Betriebswirtschaftliche Analyse des Einsatzes von bovinem Somatotropin in der Milchkuhhaltung* (Instut fur Landwirtschaftliche Betriebs- und Arbeitslehre, Universitat Kiel).

Lenoir, J. and Schockmel, L.R. (1987) 'Composition and processing characteristics of milk from cows in the French clinical trial', *Monsanto Technical Report* MLL 90311.

Ludri, R.S. *et al.* (1988) 'Bovine somatotropin in buffaloes', *Veterinary Record*, vol. 122, p. 495.

Machlin, L.J. (1973) 'Effect of growth hormone on milk production and feed utilization in dairy cows', *Journal of Dairy Science*, vol. 63, p. 575.

McNally, R. and Wheale, P.R. (1988) 'Milking the profits', *Science for People*, Issue 67, Summer, pp. 8–9.

Mix, L.S. (1987) 'Potential impact of the growth hormone and other technology on the US dairy industry by the year 2000', *Journal of Dairy Science*, vol. 70, pp. 487–97.

Moser, A.B. *et al.* (1987) 'A new dietary therapy for adrenoleukodystrophy: biochemical and preliminary clinical results in thirty-six patients', *Neurology*, vol. 21, no. 3, March, pp. 240–9.

Moser, H.W. *et al.* (1981) 'Adrenoleukodystrophy: Increased plasma content of saturated very long chain fatty acids', *Neurology*, vol. 31, no. 10, October, pp. 1,241–9.

Mouchet, C. (1987) *Consequences Economiques de l'emploi de la BST dans la production laitiere en France* (Ecoles Nationale Superieure Agronomique de Rennes).

Oldenbroek, J.K. *et al.* (1987) 'The effect of treatment of dairy cows of different breeds with recombinantly derived somatotropin in a sustained delivery vehicle', 38th Annual Meeting of the European Association of Animal Production, Lisbon, Portugal.

Omnibus Survey on Milk Consumption, Bovine Somatotropin and Trusted Sources of Information of Consumers (1988) (commissioned by Monsanto PLC) (London: British Market Research Bureau Ltd.).

Peel, C.J. *et al.* (1981) 'Effect of exogenous growth hormone on lactational performance in high yielding dairy cows', *Journal of Nutrition*, vol. 111, p. 1,662.

Peel, C.J. *et al.* (1982) 'Lactational response to exogenous growth hormone and abomasal infusion of a glucose-sodium caseinate mixture in high yielding dairy cows', *Journal of Nutrition*, vol. 112, pp. 1,770–8.

Peel, C.J. *et al.* (1985) 'The effects of long-term administration of bovine growth hormone on the lactational performance of identical-twin dairy cows' *Journal of Animal Production*, vol 41, p. 135.

Pell, J.M. *et al.* (1987) 'Effect of growth hormone on IGF-1 concentrations, body composition and growth in lambs', *Journal of Endocrinology*, vol. 112, March supplement, abstract 63.

Pell, A.N. *et al.* (1988) 'Responses of Jersey cows to treatment with sometribove, USAN (recombinant methionyl bovine somatotropin) in a prolonged release system', *Journal of Dairy Science*, vol. 71 (suppl. 1), p. 206 (abstr.).

Phipps, R.H. (1987) 'The use of prolonged release bovine somatotropin in milk production', paper available from author at AFRC Institute for Grassland and Animal Production, Church Lane, Shinfield, Reading, RG2 9AQ.

Phipps, R.K. (1988) 'The use of prolonged release bovine somatotropin in milk production', *Bulletin 228* (Brussels, Belgium: International Dairy Federation).

Poole, A.H. (1987) *The Potential Use of Bovine Somatotropins on Dairy Farms: The Implication of BST on the Structure and Size of Dairy Farming in England and Wales in 1994/95* (Farm Management Service Information Unit, Milk Marketing Board).

Popper, K.R. (1969) *Conjectures and Refutations* (London: Routledge & Kegan Paul).

Prosser, C.G. (1988) 'Bovine somatotropin and milk composition', (letter), *Lancet*, vol. 2, p. 1,201.

Prosser, C.G. *et al.* (1988) 'Increased secretion of insulin-like growth factor 1 into milk of cows treated with recombinantly derived bovine growth hormone', *Journal of Dairy Research* (forthcoming).

Schulman, S. (1989) 'US opposition to milk hormone', *Nature*, vol. 340, p. 667.

Selye, H. (1976) *The Stress of Life* (New York: McGraw-Hill).

Soderholm, C.G. *et al.* (1988) 'Effects of recombinant bovine somatotropin on milk production, body composition and physiological parameters', *Journal of Dairy Science*, vol. 71, p. 355.

Le Treut, J.H. and Desnouveaux, R. (1987) 'Checking of processing quality of milk coming from BST-treated cows: Experimental manufacturing of soft cheese (Camembert type)' *Monsanto Technical Report* MLL 90348.

Torkelson, A.R. *et al.* (1987) 'Radioimmunoassay of somatotropin in milk from cows administered recombinant bovine somatotropin', *Journal of Dairy Science*, vol. 70, Suppl. 1, p. 146.

USDA (1987) *BST and the Dairy Industry: A National, Regional, and Farm-Level Analysis*, (Economic Research Service, USDA), October.

Whitaker, E.A. *et al.* (1988) 'Health, welfare and fertility implications of the use of bovine somatotrophin in dairy cattle', *The Veterinary Record*, 21 May, pp. 503–5.

Young, F.G. (1947) 'Experimental stimulation (galactopoesis) of lactation', *British Medical Bulletin*, vol. 5, p. 155.

Part III
Genetic Engineering and the Environment

11 Introduction III

Dr PETER WHEALE and RUTH McNALLY

We are entering a new era in which genetically engineered organisms are being designed for deliberate release into the environment.

These new genetically engineered organisms include microbes for environmental control and resource-recovery. For example microbes are being genetically engineered to improve their ability to detoxify waste, digest oil slicks, leach metal and to recover oil from oil shale and tar sands. Genetically engineered organisms are also being designed for use in agriculture in the hope of increasing food production in the Third World and reducing the requirements for environmentally harmful agrichemicals. For example, scientists are trying to engineer nitrogen-fixing crop plants that will reduce the need for artificial fertiliser – an increasingly expensive input into the production of food, and a cause of pollution in the waterways. They are also attempting to make crop plants which have added genes for pest-resistance in the hope of reducing the use of chemicals to control pests, such as insects or fungi.

The work on genetically engineered live viral pesticides conducted at the Institute of Virology (IoV) in Oxford is regarded by many scientists as an exemplary research programme. In Chapter 12 Dr David Bishop, Director of the IoV in Oxford, describes the environmental testing of three different forms of a genetically engineered viral insecticide in four controlled release experiments. The objective of these studies is to assess the efficacy and environmental safety of genetically engineered viruses designed for use as alternatives to chemical pesticides.

Between 1983 and 1989 there were about 100 experimental releases of genetically engineered organisms worldwide. These experiments were designed to assess the efficacy of, and environmental risks associated with, the release of genetically engineered bacteria, plants and viruses into the environment and they pave the way for the commercialisation of the use of genetically engineered organisms in the environment.

Ecological Concerns

The prospect of the routine release of genetically engineered organisms into the environment has aroused the concern of environmentalists. The dynamic nature of genomes, together with the inherent mobility of the vectors employed in genetic engineering and the predisposition of the cells of animals, plants and microbes to incorporate 'foreign' DNA, give rise to justifiable concern about the ultimate destination of engineered genes when genetically engineered organisms are released into the environment (see Wheale and McNally 1988, Chapter 4).

In Chapter 13 Andrew Lees, a field biologist working as Water Pollution and Toxics Campaigner for Friends of the Earth, is critical of the development of genetically engineered microbial pesticides, in particular the work on viral pesticides being conducted by the IoV in Oxford.

Lees is doubtful that the use of genetically engineered organisms in agriculture will diminish the use of chemicals on the land. A number of agrichemical companies are attempting to genetically engineer crop plants which are resistant to their particular brands of chemical herbicides (weed killers). It is feared that genetically engineered herbicide-resistant plants will provide an incentive for farmers to increase the use of brand-associated chemical herbicides on their crops (see Doyle 1985).

The environmental use of genetically engineered organisms could inadvertently cause pestilence or disease. Microbial pesticides could be pathogenic to non-target species, for example, or the genes for herbicide-resistance, disease-resistance, pest-resistance, drought-resistance or nitrogen-fixation could be transferred from genetically engineered crop plants to related weed species (see Ellstrand 1988; Williamson 1988). Powerful chemicals would probably be needed to control these inadvertently created pathogens or 'super weeds', if they could be controlled at all.

Bishop acknowledges that the release of genetically engineered organisms into the environment presents potential hazards and that it is important to proceed with caution. He describes the extensive programme of environmental risk assessment which the IoV has undertaken prior to each of their planned release experiments of genetically engineered viral pesticides. These include host specificity tests and assays of the genetic stability and environmental persistence of the genetically engineered viruses. Recognising the potential for microbial pesticides to become pests themselves were they to persist in the environment, the IoV genetically engineered a 'crippled' version of the virus, whose persistence in the environment is claimed to be much diminished. However, in 'Discussion III' Bishop acknowledges that for commercial applications it may be desirable to use a fully viable viral insecticide rather than a 'crippled' one, to increase the period that the insecticide would be active in the environment.

Lees believes that the cautious approach of the IoV, laudable though it may be, will provide a 'Trojan horse' for the genetic engineering industry because it will mollify public concern over the deliberate release of genetically engineered organisms. He questions whether commercial firms will be as cautious in their release experiments as the IoV. He also expresses his lack of confidence in present-day ecology to predict the environmental effects of releasing genetically engineered organisms into the environment. This view, which is shared by many ecologists (see, for example, Simonsen and Levin 1988; ENDS 1988; Wheale and McNally 1988, Chapter 7), has led to demands for a substantial increase in public funding of research into ecology and evolution, particularly of microbes (see, for example, Regal *et al.* 1989; Rose 1989).

Regulation and Control

At the present time the regulatory provision within the EC for the deliberate release of genetically engineered organisms is at a national level and varied, ranging from Denmark, where there is a specific law – the Environment and Gene Technology Act (1986), to the UK, where genetic manipulation regulations have been appended to existing legislation for worker health and safety (see Chapter 23), through to Greece, Ireland, Italy, Portugal and Spain where there are no specific regulations for the release of genetically engineered organisms.

The lack of regulatory harmonisation within the EC is of concern both to industrialists and environmentalists. On the one hand, industrialists are concerned that such disparity may create unequal conditions of competition and thus directly affect the functioning of the Common Market. On the other hand, genetically engineered organisms do not recognise national frontiers and nothing short of Community-wide regulation can offer the necessary protection against the hazards of releasing them.

In 1988 the European Commission proposed an EC directive (EC law) to harmonise the regulations governing applications to release genetically engineered organisms into the environment (European Commission 1988). If this new directive is accepted by the European Council of Ministers, each Member State will have to bring into force the laws, regulations and administrative provisions necessary to comply with it within a period of 18 months. The British government is planning to incorporate the requirements of this proposed Directive into its environmental protection legislation, the so-called 'green' Bill introduced into Parliament in December 1989 (HC Bill 14 1989).

The draft directive on the environmental release of genetically engineered organisms requires that each EC Member State should appoint a national Competent Authority or Competent Authorities, which shall be responsible for carrying out the provisions of the directive. Before conducting a release, the person responsible for it shall submit a notification to the Competent Authority of his or her Member State, including a detailed risk assessment which identifies the possible hazards of the release.

Many ecologists consider that the potential for environmental damage arising from the release of genetically engineered organisms is great, that our theoretical knowledge about ecology and evolution and our experience of releasing organisms are limited, and that *a priori* knowledge is insufficient to predict the risks (see, for example, Simonsen and Levin 1988). Such views have prompted a number of environmental pressure groups to propose that there should be a moratorium on the release of pathogenic organisms and on the commercialisation of products consisting of, or containing, genetically engineered organisms until further research on risk-assessment has been undertaken (see 'Conference Resolutions'; Bundestag Document 1987; Gen-ethic Network 1989; Rose 1989; Schmid 1989).

These widely-held reservations about the environmental wisdom of releasing genetically engineered organisms are not reflected in the draft directive which does not prohibit any class of release as being potentially too risky to be undertaken. It would appear that those who express most concern about the potential risks of the release of genetically engineered organisms into the environment have been effectively marginalised. However, the environmental issues are not trivial and the dispute is not marginal; an amendment which proposed a five-year partial moratorium on environmental releases of products consisting of, or containing, genetically engineered organisms was defeated by only one vote when the draft directive was voted upon in the European Parliament in June 1989 (see Schmid 1989; Dickman 1989).

The proposed EC law on environmental releases does not require that national Competent Authorities be placed under legal obligation to supply public information, nor does it provide for public participation in the decision-making process (European Commission 1988). This failure to make provision for public information on specific proposals to release genetically engineered organisms into the environment contrasts with the recommendations of the Royal Commission on Environmental Pollution (RCEP)

(1989, p. 95) in the UK, the Committee on the Environment, Public Health and Consumer Protection of the European Parliament (Schmid 1989, pp. 8, 12, 29) and the opinion of a large number of non-governmental organisations in the EC, the USA, Japan and several Third World countries (Gen-ethic Network 1989; Rose 1989). A number of environmental bodies have also expressed the opinion that representatives of the public should be permitted to participate in the decision-making process in this area (RCEP 1989, p. 95; Gen-ethic Network 1989; UK Genetics Forum 1989). If the public is to have confidence in the regulatory process, then the regulatory process should provide for public access to information concerning risk assessment procedures and should permit the participation of those who represent a cross-section of public interests. Without the right to participate, the public will continue to be concerned that genetic engineering is being regulated in the interests of industry, and that those interests are not necessarily consistent with environmental welfare.

Neither the draft EC directive (Article 8) nor the 'Co-ordinated Framework for the Regulation of Biotechnology' in the USA (OSTP 1986) propose new laws for the regulation of the products of genetic engineering. Instead they propose that genetically engineered organisms should be regulated under existing product laws. This principle of 'no new legislation' is also found in the influential report of the Organisation for Economic Development and Co-operation (OECD) on the use of genetically engineered organisms in industry and the environment (OECD 1986, p. 41; see also Chapter 23). It is a principle which is shared by industrial interest groups in the USA, for example, the Industrial Biotechnology Association (IBA) (see Milewski 1986, p. 32) and in the EC (see Poole *et al.* 1988). Interestingly, whilst the major European industrial interest group – the European Biotechnology Coordinating Group (EBCG) – was consulted during the drafting of the directive, employee representatives and environmental pressure groups were not.

Edward Lee Rogers, legal adviser to Jeremy Rifkin of the Foundation on Economic Trends, does not believe that existing statutes which were enacted before the advent of the new techniques of genetic engineering address its unique risks adequately. In Chapter 14 Rogers points out his concerns about regulatory gaps and overlaps which have emerged in the co-ordinated framework in the USA. Concern has been expressed that a similar situation is likely to arise in the EC, especially considering that the majority of product laws have not been subjected to scrutiny for their suitability to regulate genetically engineered organisms (see RCEP 1989,

p. 70; Schmid 1989, p. 29). Rogers reports that because of the serious gaps and inconsistencies within the co-ordinated framework in the USA, individual states are beginning to introduce additional legislation to protect the environment. Equivalent action on a Member State level in the EC may not be possible because the draft directive is primarily a measure to regulate trade. This will limit the unilateral steps that individual Member States can take to adopt stricter regulations (see also Wheale and McNally 1990).

In 1989 the European Parliament's Committee on the Environment, Public Health and Consumer Protection published a report criticising the proposed directive (Schmid 1989). In the opinion of this committee, the draft directive could have done more to protect the environment by introducing measures to enforce adherence to its provisions and to make those releasing genetically engineered organisms responsible for any adverse environmental consequences arising as a result of their releases. The committee suggested that responsibility for environmental damage following the deliberate release of genetically engineered organisms should be under the legal principle of 'strict' (or 'absolute') liability, whereby any individual or organisation claiming for damages caused by another party does not have to prove that the other party acted negligently in order to claim damages, but merely to show that the damage was caused by the actions, activities or products of the other party. The committee also suggested that a Competent Authority should not authorise a deliberate release unless full and appropriate insurance coverage for the particular deliberate release has been provided by the applicant (Schmid 1989).

A system of responsibility and accountability for genetically engineered organisms is dependent upon a reliable system for monitoring them in the environment. The IoV programme of research at Oxford has demonstrated that organisms which are released into the environment can be labelled by inserting a small unique piece of DNA into their genetic material. This piece of DNA, called a 'genetic marker', enables the positive identification of released organisms. In written evidence submitted to the Royal Commission on Environmental Pollution (RCEP) the authors recommended a system of environmental monitoring of genetically engineered organisms based on the use of genetic markers (RCEP 1989, p. 46; Wheale and McNally 1988, pp. 128, 132–4). As a condition of obtaining permission to undertake a release experiment, the unique genetic marker could be publicly registered with the appropriate regulatory authority. Were the genetically engineered organism to cause environmental damage, the genetic marker would enable the party responsible for releasing it to be identified and held accountable.

A public register of genetic markers would certainly encourage organisations who wish to license and sell living genetically engineered organisms as products to act responsibly. First, it would be in their own interest to develop genetically engineered organisms that were genetically 'stable' to safeguard against the possible transfer of their unique genetic marker to other organisms for whose environmental damage they could be held liable. Secondly, it might encourage more organisations to develop genetically engineered organisms of limited environmental persistence to minimise the risk of undesirable side-effects. Thirdly, it would encourage organisations to be more vigilant in their pre-release risk assessments, especially with respect to the undesirable long-term effects of their genetically engineered organisms. Moreover, such organisations may consider that it is in their own interest to ensure that those who purchase their products are properly instructed in their safe use.

The clear division in the EC between those supporting the draft directive on the deliberate release of genetically engineered organisms, and those opposing many aspects of it and demanding substantial amendments to it, is indicative of a schism between, on the one side, scientists associated with genetic engineering together with their industrial supporters, and on the other, those primarily concerned with protecting the environment.

Third World Perspective

The so-called 'Green Revolution' of the 1960s was the replacement of traditional crop plants by high-yielding varieties. Whilst solving some of the food production problems of the Third World, the agrichemical and irrigation requirements of these high-yielding varieties created many problems, including adverse ecological effects and increased expenditure on food production inputs. The novel plants and seeds developed in the advanced industrial countries are sold to Third World countries where they tend to displace indigenous varieties (see, for example, Doyle 1985; Wheale 1986).

The close involvement of transnational agripetrochemical companies in the research and development of genetically engineered crop plants – herbicide-resistant crops, for example – together with the extended intellectual property rights which facilitate the appropriation of the world's stock of seeds and germplasm (see Chapter 1) give little assurance to Third World countries that the Second Green Revolution which genetic engineering promises will succeed where the First Green Revolution failed (see Wheale and McNally 1988, pp. 157–61).

Patents and plant breeders' rights favour scientific and technological invention over social and cultural invention and environmental conservation. Inventive stewardship of natural genetic resources remains unprotected by patent law whereas technological intervention and control is rewarded by economic monopoly. In such a system, the patenting of life-forms favours the commercial interests of the advanced industrial countries over the interests of agronomy, ecology and conservation.

The agrigenetic engineering revolution begins by appropriating the genetic resources of the Third World, where 95 per cent of the world's plants originated. However, the Third World has neither the international legal right nor the technological or economic means to challenge the intellectual property rights claimed by transnational companies over their genetic resources. The extension of patent law to living organisms extends the exploitative, manipulative and monopolistic worldview to the biosphere (see Juma 1989).

12 Genetically Engineered Insecticides: The Development of Environmentally Acceptable Alternatives to Chemical Insecticides

Dr DAVID BISHOP

The use of genetically engineered organisms in the environment represents a new era of scientific endeavour. It is therefore important to develop the subject systematically and cautiously, taking into consideration the consequences, particularly the risks, of any prospective introduction. Indeed it would be irresponsible to do otherwise.

The objective of the programme of research and development of genetically engineered baculovirus insecticides that is underway at the Natural Environment Research Council (NERC) Institute of Virology (IoV) at Oxford is to assess the consequences of their deliberate release into the environment.

Baculovirus Insecticides

Baculoviruses are a particular group of viruses. Members of the Baculoviridae family are unique because they only infect arthropods, for example, insects and crustaceans. No member of this family of viruses infects any other form of life.

Baculoviruses are pathogenic (cause diseases) in certain insect species. Studies of naturally occurring epizootics of baculovirus-induced disease have also shown that the effect of baculoviruses is limited to certain insect species. (An epizootic is a disease affecting a large number of insects simultaneously, corresponding to an epidemic in humans.) The principal target of baculoviruses is the larval stage of the insect host. For butterflies and moths this is the caterpillar.

Naturally occurring baculoviruses have been used as biological pesticides to control insect pests since the last century. Several hundred baculoviruses have been isolated and more than a dozen have been employed commercially to control insect pests. Reports by Podgewait (1985) and Entwistle and Evans (1985) list some of the baculoviruses which have been used for pest control.

Unlike many chemical insecticides, baculovirus insecticides affect only a few species of insect – the so-called 'permissive hosts'. Baculovirus insecticides have no harmful effect on insects other than their permissive hosts, nor do they harm other invertebrates, plants, or vertebrates, and they do not pollute the environment in terms of causing adverse reactions in soil or water. The environmental safety of baculoviruses is a matter of record, and a major factor when considering their further development as pesticides through genetic engineering. This record of safety has been confirmed by experience gained over many decades involving the use of naturally occurring baculovirus insecticides in agriculture and forestry.

A joint meeting of the World Health Organisation (WHO) and the Food and Agriculture Organisation (FAO) on insect viruses endorsed the potential of baculoviruses as pest control agents (WHO/FAO 1973). The report noted that, in addition to their specific host ranges, baculoviruses exhibit good storage properties, are safe to handle, are relatively easy to produce and are widely distributed in nature, particularly among insects.

Infection Cycle of the Nuclear Polyhedrosis Viruses (NPVs)
Nuclear polyhedrosis viruses (NPVs) are a sub-group of the baculoviruses. NPVs have a gene for a protein called the polyhedrin protein. This protein forms a multi-sided, crystalline structure in the shape of a polyhedron in which many NPVs are enclosed, forming what is known as a polyhedrin inclusion body (PIB).

Figure 12.1 is a schematic representation of the infection cycle of NPVs. The cycle starts when caterpillars ingest PIBs with their food. When the PIBs reach the alkaline environment of the caterpillar's gut, the polyhedrin protein is digested by enzymes, releasing the viruses which infect cells in the lining of the caterpillar's gut where they replicate. The newly-replicated viruses then spread the infection to other cells of the caterpillar. Late in the infection cycle polyhedrin protein is made and viruses are packaged into PIBs. The cells and tissues of the caterpillar break down, killing the caterpillar and releasing the PIBs. Up to 10^9 PIBs may be present in a single caterpillar corpse.

Caterpillars infected with NPVs commonly exhibit behavioural abnormalities which may be of benefit to the subsequent spread of

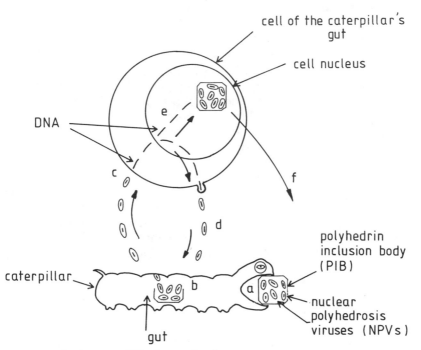

a) Caterpillar ingests a PIB when eating leaves.
b) The polyhedrin protein is digested in the caterpillar's gut, releasing the viruses.
c) The viruses infect the gut cells where they replicate.
d) Newly-replicated viruses released from the gut cells infect other caterpillar cells.
e) Late in the infection cycle, polyhedrin protein is made and viruses are packaged into it to form PIBs.
f) The infected cells break down releasing PIBs which are eaten by other caterpillars.

Figure 12.1:
Schematic representation of the infection cycle of the nuclear polyhedrosis viruses (NPVs)

virus to other caterpillars. For example infected caterpillars may ascend to the tips of plants prior to death, thereby facilitating the distribution of PIBs over the foliage below.

PIBs released from decaying caterpillar corpses can be spread by physical forces, for example, rain splash. They may also be distributed passively by animals that eat the remains of the caterpillars and subsequently defaecate the PIBs at other sites. When birds, rodents and beneficial insects, such as beetles, ingest PIBs they do not become infected with the viruses because the viruses pass straight through their intestines and are deposited in their faeces.

IoV Research Programme on Genetically Engineered Baculovirus

The objective of the IoV research programme on genetically engineered baculovirus insecticides is to improve their speed of action. This is desirable because during its normal infection cycle a baculovirus undergoes several rounds of replication (see Figure 12.1). These rounds of replication take time – up to several days or weeks depending on the virus, the host and the environmental conditions, for example, temperature. By contrast, most chemical insecticides act quickly, killing the target insect (and often other insects, including beneficial insects) in a matter of hours. Using genetic engineering procedures it should be possible to incorporate foreign genes, for example insect hormone genes, into the baculoviral genome (genetic material) which will reduce the time the viruses take to kill the target caterpillars.

Since 1986, when the IoV field trials of genetically engineered baculovirus insecticides were initiated, three forms of a genetically altered baculovirus have been developed and released in a field facility. The facility used for the studies is on open arable land at the Oxford University Field Station at Wytham, Oxfordshire.

The IoV programme of research has used a strain of baculovirus called *Autographa californica* nuclear polyhedrosis virus (AcNPV). The first IoV environmental release involved a genetically marked AcNPV (see Bishop 1986). The second and third releases by the IoV, undertaken in 1987 and 1988 respectively, endeavoured to determine whether it was possible to genetically engineer a marked and 'crippled' AcNPV insecticide which would be degraded quickly in the environment. Such a virus should be even safer than the natural virus and would be a suitable substrate for further engineering. The virus was 'crippled' by the removal of the gene coding for the polyhedrin protein, thereby creating a 'polyhedrin-negative' AcNPV which was rapidly degraded in the environment by ultraviolet light.

The fourth release was undertaken in 1988 and involved a 'crippled' AcNPV that contained a 'junk' (beta-galactosidase) gene as a marker. Unlike the markers in the earlier releases, which were latent pieces of DNA, the marker DNA in this experiment was expressed by the virus as a protein, and is known as a 'phenotypic' marker.

The First Release

Pre-release Risk Assessment
The use in the environment of a genetically engineered organism represents a new era of scientific endeavour, and therefore it is important to develop the subject systematically and cautiously. As mentioned above, it would be irresponsible to construct and release into the environment a genetically engineered virus that expresses a foreign gene without a thorough understanding of the possible consequences. The programme of research and development that is underway at the NERC IoV at Oxford is therefore primarily concerned with the assessment of risks associated with the introduction of genetically engineered baculovirus into a natural habitat.

There is one point which should be made concerning the suitability of using baculoviruses as a model for risk assessment analyses of the deliberate release of genetically engineered organisms. Baculoviruses, like all viruses, are capable of reproduction only when inside a living cell of their permissive host species. Outside of such cells, viruses remain inert in the environment, unable to replicate, and eventually degrade. In this regard viruses differ from most bacteria and other free-living micro-organisms and are therefore not models for assessing the risks involved with genetically engineered free-living organisms, such as bacteria. However, this property makes viruses excellent subjects with which to conduct a risk assessment programme since their infection requirement (presence of a permissive host species) allows restrictions that limit risk to be imposed on the study.

Many of the risks associated with the introduction into the environment of a genetically engineered organism can be assessed in the laboratory before undertaking a release. The task of pre-release risk assessment, however, is not trivial, and often involves several years of investigation.

For each environment release experiment we have conducted, we have performed the following pre-release risk assessments: the construction of the candidate virus insecticide; verification at the nucleotide level of the precise genetic change; laboratory analyses of the phenotype and genetic stability (mutability) of the virus; host range and genetic compatibility determinations, and analyses of the physical stability of the virus in simulated systems (plant surfaces and soil). Below I describe the pre-release risk assessments undertaken for the first release in 1986.

Construction of a Marked AcNPV
The first release, undertaken in 1986, involved a genetically marked AcNPV that was otherwise identical in every way to the parent AcNPV (Bishop 1986).

The virus was genetically engineered so that it could be distinguished from all other baculoviruses, including the unmodified parental AcNPV. The distinguishing feature of the virus was that a piece of DNA was inserted into its genome (genetic material). The only purpose of this piece of DNA, known as the genetic marker, was to serve as a flag so that the marked virus could be detected by a simple procedure.

When inserted the genetic marker was located so that it neither added to the constitution or synthesis of any of the natural viral gene products nor detracted from them. This was verified by comparing the infectivity of marked viruses and the proteins synthesised by them with those of the unmarked parental viruses.

Host Range

A major issue in terms of risk is whether insects which were not susceptible to infection by the parental viral strain become liable to infection by the engineered strain. This is of particular concern where there is a rich insect fauna at the site chosen for the release. To address this issue, extensive host range analyses have been undertaken in the laboratory for each of the genetically engineered viruses we have prepared, prior to seeking permission for field trials.

To assess the effect of the presence of the genetic marker on the host range of the virus, some 90 species of insect were analysed for sensitivity to the marked virus and to the parent virus used to obtain it. The insect species chosen were selected in consultation with the Nature Conservancy Council (NCC) and focused particularly on UK butterflies and moths.

The host range tests involved collecting the adult insect species, allowing them to lay eggs (or collecting eggs from native plants), surface-sterilising the eggs using formalin as described by Hunter *et al.* (1984), and permitting the larvae to hatch. Larvae were then infected with virus by allowing them to ingest known numbers of PIBs with their preferred diet. Control larvae, not infected with virus, were treated similarly.

For convenience of handling, usually second or early third instar (developmental stage between the second and third moults) larvae were employed in batches of 10–30 (depending on availability) per dose of virus (10^{2-6} PIBs per individual). When limited numbers of larvae were available only the higher doses of virus (10^{4-6}) were used.

The larvae were fed and kept until they pupated (formed a chrysalis) or died, and then analysed for virus infection as described by Wigley (1980). The presence of virus was verified using procedures (see Possee and Kelly 1988) which identified the presence of the polyhedrin gene and the genetic marker.

On occasion, pupae were allowed to develop into adults of the species, although other than observing that the adults had a normal gross morphology and, after mating, were capable of laying fertile eggs, no systematic analyses of the effects of viruses on subsequent insect generations were undertaken.

If sufficient numbers of larvae perished as a result of virus infection, the data were used to calculate the lethal dose that caused death in 50 per cent of the insects (LD50) using statistical analyses as described by Finney (1971). This was only possible in the tests performed with the cabbage looper moth *(Trichoplusia ni) (T. ni)* and the small mottled willow moth *(Spodoptera exigua) (S. exigua)*. Other species proved to be resistant to infection or were only susceptible when the highest doses of virus were administered.

The results of the host range studies are summarised below. Most of the UK insect species tested were not susceptible to either the marked or the parent virus and no difference in host range was found between the marked virus and the parent virus.

Included in the host range studies were ants, honey bees, lacewings, hoverflies and various beetles and ladybirds. As expected, none of these insects was found to be susceptible to infection by the marked or the unmarked AcNPV. Also in the list of species which were not susceptible to infection by either the marked or the unmarked virus were all 17 butterfly species tested (representing five families of butterflies), several microlepidoptera, and some 58 moth species (representing eleven families).

Caterpillars of three moth species that are not native to the UK were found to be permissive for virus infection. Similar results were obtained with these species using marked and unmarked viruses. All three moths are members of the Noctuidae family. The most sensitive species was the cabbage looper moth (LD50: 10^2 PIBs/individual). The other two species were only sensitive at doses of 10^{5-6} PIBs/individual, although insufficient caterpillars died to allow the LD50 to be calculated. One UK member of that family, the small mottled willow moth, was quite sensitive to the viruses (LD50: 10^5 PIBs/individual). Two UK members of the Sphingidae, the lime hawkmoth *(Mimas tiliae)* and the privet hawkmoth *(Sphynx ligustri)*, were also permissive, although only when infected with high doses of the viruses (10^{5-6} PIBs/individual). Such doses are probably seldom involved in the initiation and maintenance of natural epizootics (insect epidemics). Again, insufficient caterpillars died to allow the LD50 to be calculated.

Natural epizootics are usually caused by NPVs in permissive species which have low LD50 values (see, for example, Lewis *et al.* 1981; Hughes 1978). Although the small mottled willow moth was susceptible to AcNPV, the LD$_{50}$ was high. In view of this, caterpillars of the small mottled willow moth were considered to be suitable as target hosts for the field studies.

Spread of the Virus

Another element of pre-release risk assessment is the possible spread of the virus from the site of application which may be mediated by environmental conditions such as rain, or by the movement of the host species, or by other animals, for example, birds, rodents and predatory insects. Entwistle *et al.* (1983) have demonstrated the occurrence of such spread by studying both natural epizootics of viral infection and induced epizootics when viruses were applied to caterpillar infestations in the wild. However in such investigations it was never entirely possible to determine whether the virus recovered from the environment originated from the applied virus, or whether it was from a naturally occurring population of the same virus.

It is necessary when assessing the spread of viruses which have been deliberately released into the environment to be able unequivocally to distinguish them from other viruses. It was in order to address this need that the genetically marked virus was developed.

Gene Transfer

Another potential risk consequent to the release into the environment of a genetically engineered virus is the possible transfer of genetic information.

Gene transfer from a virus into dead cells or non-germline (non-sex) cells would be an event without consequence. The transfer of DNA from cells of a host species to a virus could occur (albeit infrequently) and could alter the properties of the virus.

Although gene transfer is frequent and well-documented among bacteria (for example, via plasmids and conjugation), gene transfer is probably a low-frequency event among baculoviruses, with the exception of that between closely related baculoviruses in ecologically intimate circumstances. Gene exchange between genetically compatible baculoviruses may occur when two closely related baculoviruses co-infect the same host. Studies indicate that the rate of gene exchange between closely related viruses could be greater than 1 per cent. The likelihood of gene exchange occurring between such viruses depends on whether the target host is already infected with a related virus, or whether another virus co-infects the same target at the same time.

If gene transfer occurs naturally, then the question is whether it occurs more frequently in the genetically engineered virus than in the parent virus, and whether the presence of the genetic marker induces the transfer of particular DNA sequences.

So far in our studies we have not been able to identify a UK baculovirus that is closely related to AcNPV, or that is genetically compatible with AcNPV. Data we have obtained from studies with a distantly related baculovirus indicate that if gene transfer does occur it only occurs at frequencies lower than 10^{-7}; to date, no such gene transfer has been observed in our studies.

Genetic Stability
Another consideration is the question of whether the introduced gene induces genetic instability in the engineered virus. This can be investigated in the laboratory by analysing the viruses for signs of change in their genetic material after they have undergone multiple rounds of replication in (been 'passaged' in) either living caterpillars or in caterpillar cells in culture.

In order to compare the genetic stability of the marked virus with the genetic stability of the parent virus, both were passaged for some 50 cycles of replication in the cells of fall armyworm (*Spodoptera frugiperda*) (see also Brown and Faulkner 1977). Virus was also passaged in cabbage looper moth caterpillars.

These studies detected no difference in genetic stability between the marked and unmarked viruses. Following repetitive passage of the virus both in cell culture and in insects no alteration in the viral genome could be detected through analyses involving a variety of restriction enzymes, or by protein analyses. Neither were any new gene sequences (such as host DNA) identified in the passaged virus. Sequence analysis indicated that the genetic marker remained stable, and it is predicted that it should remain so indefinitely.

In summary, genetic instability has not been detected in the genetically engineered baculoviruses that have been produced and analysed in our laboratory.

Physical Stability
Prior to the field release the physical stability of the marked virus was determined by the following procedures. Soil was collected from the proposed field site and steam-sterilised. PIBs from either the marked or parent virus stocks were mixed with the soil which was then left in the laboratory at room temperature (about 18°C) for two weeks. Samples were taken throughout this period from which PIBs were recovered using differential centrifugation. The recovered PIBs were tested for residual viral infectivity using cabbage looper moth caterpillars as biological indicators. No significant decline in infectivity for either the marked or unmarked virus was detected during the two week period of the experiment.

These studies in the laboratory indicated that the physical stability of the marked virus was not affected by contact with soil over a period of two weeks; the virus retained full infectivity throughout this time. In view of these results, it was considered inevitable that soil at the field site would become contaminated with virus. This was found to be the case, and consequently the site was disinfected at the termination of the study.

The host range and physical stability of the marked virus were also as expected. The genetic marker did not affect the replication of the virus in cell culture or in insects. Furthermore, its design and site of insertion was such that it could not and, from the studies conducted, did not alter the synthesis of any other virus gene product.

Independent Evaluation

Once the pre-release risk assessment data were obtained and evaluated, field trials were required to confirm the results and establish the validity of the predictions. Approval for the release of the marked virus used was sought in 1985 and in 1986 from a number of regulatory authorities and interested parties.

The pre-release risk assessment data were submitted to all the relevant regulatory authorities for independent evaluation especially in respect of the anticipated environmental impact, taking into account the characteristics of the site proposed for the release, the insect fauna of the area, the possibility for spread beyond the immediate ecosystem, and the proposed physical constraints imposed at the release site. Following consultation on the design of the field site and host range tests, permission was given, in the form of licences, to undertake the releases described below under certain prescribed conditions (for example, site arrangements, procedures to be followed for the introduction of the engineered virus and disinfection of the site after use). Since baculoviruses are insecticides, the licences were issued by the UK Ministry of Agriculture, Fisheries and Food (MAFF) under the Food and Environment Protection Act (1985) and the Regulation of the Control of Pesticides Regulations (1986).

Primary evaluation of the proposals was made by the Health and Safety Executive (HSE) Advisory Committee on Genetic Manipulation (ACGM), in consultation with the MAFF, the Department of the Environment (DoE) and the NCC. In addition other interested parties were informed including the NERC senior management, the owners of the field site (Oxford University), senior authorities at Oxford University, the Oxford University Safety Officer, the Oxford HSE factory inspector, the Vale of the

White Horse Environmental Health Officer, the Environmental Services Committee of the Vale of the White Horse District Council and agencies including Friends of the Earth (FoE).

Press coverage (national and local papers, including, for the prospective 1988 releases, a 'Public Notice' placed in local papers and displayed in the local Post Office), as well as radio and television interviews also aided in informing the general public of the proposed studies.

The Field Site
For the field studies, it was decided to use caterpillars of the small mottled willow moth as the target pest. This is a noctuid (night) moth with practically a worldwide distribution. It is a seasonal immigrant to the UK. It eats a range of plants (herbaceous dicotyledons) including sugar beet, and has a wide distribution in communities of low-growing plants, both wild and cultivated. Caterpillars of this moth pass through five instars (effectively, five moults).

The site chosen and used for the release of the marked virus (and the other genetically engineered baculoviruses in 1987 and 1988) was situated in a field of light loam soil bounded by agricultural land at the Oxford University field station at Wytham, Oxfordshire. The area has been extensively studied with regard to flora and fauna for many decades. Moth species native to the region were collected at night using two ultraviolet light traps. The moths were trapped throughout the late spring, summer and autumn of 1986 and 1987.

The inability of a virus to replicate autonomously outside of the cells of its permissive host limits its potential to escape from the field facility and multiply. To further restrict any unwanted impact of the genetically engineered viruses on the environment, the IoV field trials were undertaken during the autumn when the natural permissive insect host of the baculovirus was not present in the environment. Caterpillars for the field trials were supplied from the laboratory.

The nearest human population to the field site is some 0.25 miles distant. The facility consisted of a 10 metre square, netted, insect-proof enclosure that was surrounded by a 2 metre high wire fence to prevent access by large animals (see Figure 12.2a).

The netted facility restricted insects, spiders, aphids, centipedes and millipedes from either entering or leaving the enclosure. No moth or butterfly species were detected in the facility during the course of the study, although some beetles and spiders appeared within the netted enclosure – presumably they developed from eggs or other life-stages dormant in the soil.

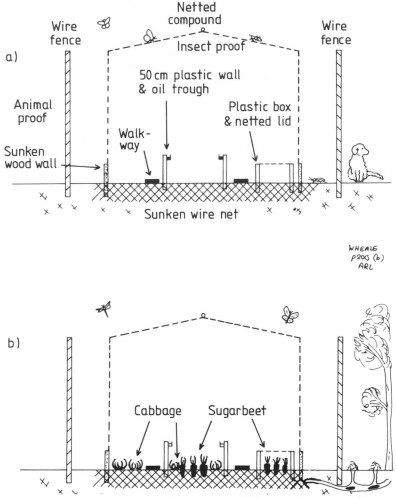

a) The outer, large animal-proof wire fence and the inner insect-proof, mole-proof and rodent-proof netted compound (dashed line). The central area of the compound contained the 2x2m plot used for the release. This is bounded by a plastic wall with an upper oil trough to prevent the escape of the caterpillars.

b) Where the sugarbeet and cabbages were planted.

Figure 12.2:
Schematic representation of the field facility used for the release of genetically engineered *Autographa californica* nuclear polyhedrosis virus (AcNPV)

The field facility was designed to restrict the escape of the infected caterpillars and to limit entry of rodents, moles or larger animals (see Figure 12.2a). Wire netting was inserted into the ground all around the enclosure to prevent moles and rodents from gaining entry. Many fresh mole hills were formed in the field throughout the summer and autumn of 1986, and again in 1987 and 1988; none occurred within the facility during, before or after the study.

Within the netted facility, a plot of 4 square metres, open to the air, was surrounded by a 50 centimetre high plexiglass wall with an upper, encompassing, oil trough. The plexiglass was inserted 25 centimetres into the soil to restrict the escape of the caterpillars and the entry of other arthropods or other animals. The plot was planted with sugar beet and cabbages for the virus release experiments (see Figure 12.2b).

By such physical and temporal restraints as described above, any risk to the environment as a result of the introduction of a genetically engineered virus or its spread was minimised.

The Planned Release Experiment
Eventually, after obtaining the requisite permissions (see above) the first field trials with the marked AcNPV were undertaken in September 1986.

About 200 early third instar caterpillars of the small mottled willow moth were fed overnight in the laboratory with ten LD50 PIB equivalents of the marked AcNPV incorporated into an artificial diet plug. The infected caterpillars were re-fed with diet plugs containing similar quantitites of the marked virus and, after they had eaten it all, they were taken in sealed containers to the site and placed on the sugar beet plants in the central plexiglass-enclosed area of the containment facility (see Figure 12.3a). Uninfected caterpillars were released on to nearby plants in an adjacent, totally netted enclosure. After one week none of the infected caterpillars remained alive (see Figure 12.3b). Approximately 55 dead caterpillars were evident on plant surfaces. Other dead caterpillars were observed in or on the surface of the soil.

Samples of soil and foliage (cabbages, sugar beet and chickweed that grew within the facility) were returned periodically to the laboratory at one to two week intervals for analysis for virus. These analyses were continued over a six month period until February 1987.

Virus was assayed on the recovered plants by allowing groups of second instar cabbage looper moth caterpillars to completely eat the foliage. The caterpillars were then reared on an artificial diet until pupation or death, when they were assayed for virus, as described above. As expected for plants that originally had only a

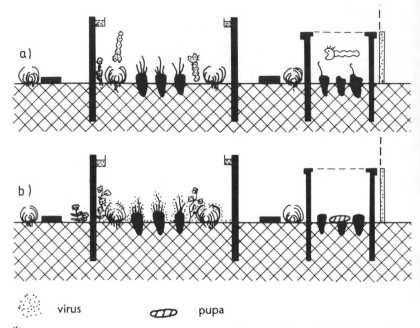

 virus   pupa

Laboratory infected caterpillars of the small mottled willow moth (*S. exigua*)

Uninfected caterpillars of the small mottled willow moth (*S. exigua*)

a)The start of the field experiment, showing infected caterpillars in the central area of the plexiglass-walled compound on the left hand side (see also Figure 12.2a) and uninfected caterpillars in the totally netted enclosure on the right hand side.

b)The end of the field experiment, showing virus in the soil and on foliage in the left hand enclosure, and uninfected pupae in the right hand enclosure.

Figure 12.3:
Schematic representation of the first Institute of Virology (IoV) field trials of genetically engineered baculoviruses

limited number of infected small mottled willow moth caterpillars on them (and consequently a limited amount of residual virus), not all the cabbage looper moth caterpillars became infected with marked virus. However, the virus was identified at every sampling period throughout the entire six months. It was identified on plants that were present in the site at the initiation of the experiment (cabbages, sugar beet), as well as on those that grew up in the facility after the small mottled willow moth caterpillars had died (for example chickweed). The amounts of virus on the infected plants or in the soil were not quantified.

The virus which was identified after replication in cabbage looper moth caterpillars had retained the genetic marker in its original form. Chickweed recovered from outside the facility and plants collected outside the enclosed area did not yield marked virus.

The presence of marked virus in soil was assayed by bringing soil samples back to the laboratory and planting cabbage seeds in them. The derived cabbage seedlings were fed to cabbage looper moth caterpillars which were then reared until pupation or death and assayed for the marked virus as described above. Again, although only about 12 per cent of these caterpillars died of virus infections, marked virus was identified by the assay procedure at each sampling period. In summary, the study demonstrated the persistence of the marked virus in the field site, as expected.

In the completely netted enclosure where uninfected caterpillars were placed, the small mottled willow moth caterpillars consumed substantial quantities of foliage for three weeks after the release. At the end of that period the caterpillars were removed, all the plants were destroyed by autoclaving (that is, by conditions of extreme heat and pressure) and the area was flooded with water to retrieve pupae. Laboratory tests did not reveal the presence of marked virus in these pupae.

In order to estimate the survival of PIBs on leaf surfaces a second experiment was conducted. In an area of the netted enclosure *separate* from the walled, central area used for the release of infected caterpillars, known concentrations of the genetically marked AcNPV (expressed as *T. ni* LD50 equivalents) were placed on the upper surfaces of cabbage leaves. Leaves from these plants were harvested at intervals and fed to second instar cabbage looper moth caterpillars in the laboratory to determine the rate of virus loss. Virus loss (and/or inactivation) was estimated from this experiment to be of the order of 10 *T. ni* LD50 equivalents of infectious virus per week. Since many times during this period of the experiment plant surfaces came into contact with rain, some loss may have been due to the virus being washed off the leaf surfaces. Such loss, if it occurred, was not measured.

Finally, at the completion of the experiment in February 1987, the field site was cleared and the foliage was removed and sterilised to destroy any remaining virus. The site was treated with 5 per cent formalin on three separate occasions, twice in February and again in March, 1987. This concentration of chemical is known to inactivate polyhedra very effectively as shown by Hunter *et al.* (1984).

Soil samples were taken before and after each formalin treatment and used in the laboratory to propagate 200–400 cabbage seedlings.

The seedlings were fed to cabbage looper moth caterpillars which were to act as a biological indicator for the presence of virus. The caterpillars were reared until pupation, or death. All caterpillars and pupae, infected or healthy, were then assayed in order to detect the presence of the marked virus as described above.

The results showed that before the first formalin treatment, infectious virus was present in the soil in amounts sufficient to infect most of the cabbage looper moth caterpillars. No attempt was made to quantify the numbers of infectious virus. After the third formalin treatment no virus was detected in the soil samples; all the cabbage looper moth caterpillars that were fed on cabbage grown on the soil sample remained healthy, pupated and developed normally. None contained any detectable quantity of the marked virus.

Summary of the Results of the Planned Release Experiment
This experiment demonstrated that caterpillars of the small mottled willow moth which were infected with genetically marked AcNPV died from the virus infection. Marked virus was recovered from plant surfaces (and the soil) over a six-month period after the release, demonstrating the resistance of the marked virus to environmental degradation. In a separate experiment as part of the field study, the rate of loss of genetically marked virus from plant surfaces was shown to be a ten-fold loss per week. Whether this was due only to the effects of ultraviolet light, or other causes of inactivation, or to washing off from the plants by rain water, is not known. Formalin treatment of the site successfully inactivated the remaining virus.

The Second and Third Releases

The impact of baculoviruses on insect populations is greatly affected by their physical stability, that is, their capacity to retain infectivity over long periods of time, and to occur in situations where they can infect a new host. Persistence of the baculovirus in the environment is necessary in order for it to infect subsequent generations of insects, generations that may not occur until the next year or subsequent years. As discussed by Carruthers *et al.* (1989), the prolonged presence of infectious baculoviruses on plant surfaces between host generations is probably a result of the release of PIBs from slowly decaying caterpillar corpses and also because of the persistence of PIBs on plants and in soil.

The polyhedrin protein of NPVs provides them with a degree of protection against ultraviolet light inactivation and probably also facilitates the retention of virus on particular plant and other surfaces. PIBs may persist for several years in the soil and can be acquired by seedlings growing in the soil, thereby bringing baculovirus into contact with the target host – the feeding larvae of certain insects. The subject of insect virus persistence in the environment has been reviewed by Evans and Harrap (1982).

Construction of a 'Crippled' and Marked AcNPV

A second release of a genetically engineered AcNPV baculovirus insecticide was undertaken in 1987. This time the polyhedrin protein gene and its controlling sequence – the transcription promoter – were removed in their entirety from the virus genome and replaced by a unique genetic marker. The removal of the polyhedrin protein gene and its transcription promoter rendered the viruses incapable of forming PIBs. Such viruses are called 'polyhedrin-negative', 'crippled', and 'self-destruct' viruses.

The genetic marker was designed so that it would neither add to, nor detract from, the expression of any viral gene product. Its sequence and position in the viral genome were verified by sequence analyses. No alteration to the phenotype of the virus was detected by protein analyses, other than the inability of the virus to form PIBs as predicted by the loss of the polyhedrin protein gene. The genetically marked and 'crippled' virus was then tested with regard to its genetic and physical stability and its host range among UK insects.

Host Range

The host range of the genetically marked and 'crippled' baculovirus was investigated as before. The analyses involved testing bees, wasps, ants, beetles, ladybirds, alder flies, lacewings, ant lions and true flies and a large selection of moths and butterflies.

Of the UK species tested, the only moths that were identified as permissive for the marked and 'crippled' virus were the small mottled willow moth (LD_{50}: 10^6 infectious virus particles, termed 'plaque-forming units' (PFU)/individual) and the lime hawkmoth and privet hawkmoth (Bishop *et al.* 1988). The latter species were only infected at doses of 10^6 PFU/individual. No other insect species among the UK fauna that were tested were found to be susceptible to infection by the marked and 'crippled' virus. For the cabbage looper moth, the LD_{50} was calculated to be 10^3 PFU/individual; higher, as in the case of the small mottled willow moth, than with the occluded AcNPVs (PIBs).

Genetic Stability

The genetic stability of the marked and 'crippled' virus was investigated following several passages of the virus in cabbage looper moth caterpillars and 25 consecutive passages in tissue culture. The marked and 'crippled' virus was found to be genetically stable through an estimated (minimum) 50 cycles of virus replication, as demonstrated by both genotype (restriction enzyme) and phenotype (protein and infectivity) analyses. No novel sequences, genome rearrangements or any other altered properties were demonstrated.

Physical Stability
Physical stability tests were undertaken as before. Soil was collected from the proposed field site and steam-sterilised. The marked and 'crippled' baculovirus was mixed with soil and left in the laboratory at room temperature (about 18°C) for two weeks. Samples were taken throughout this period, and assays indicated a 100-fold decline in virus infectivity in three days of incubation. Since prior experiments with the marked virus indicated that occluded AcNPV (PIBs) remained infective in soil, as described above, these results showed that, by comparison, the polyhedrin-negative virus was phenotypically 'crippled' as a consequence of the removal of the polyhedrin protein gene and was inactivated or degraded in soil. Similarly, a greatly increased sensitivity to ultraviolet light inactivation was demonstrated.

Planned Release Experiments
As before, after review of the laboratory data and independent evaluation, a licence was issued by the MAFF to undertake a release under the same conditions and using the same protocols and facility that were employed for the occluded, marked AcNPV described above (see Figure 12.3). The objective was to determine if the marked and 'crippled' baculovirus released from dead caterpillars persisted in the environment in soil and on plant surfaces. As before, caterpillars were infected in the laboratory and taken to the field site and released in the central enclosed area on to sugar beet. At the end of one week all the caterpillars had died. One week later, no virus capable of infecting second instar cabbage looper moth caterpillars was identified, either on the plant surfaces or in soil samples. Even the corpses of dead small mottled willow moth caterpillars did not yield infectious virus.

In summary, the data obtained with the marked and 'crippled' baculovirus (the so-called 'self-destruct' virus), demonstrated that it did not persist in the environment. To determine whether the virus could be used as a biological control agent *per se*, a third release using the same virus was undertaken in 1988 to measure virus efficacy. Virus was placed on sugar beet leaves on which small mottled willow moths had laid eggs. The plants were then transferred from the laboratory to the field facility to monitor infection and survival rates. The dose response data indicated that when the virus was used at high doses all the emerging caterpillars died from virus infection and that no infectious virus remained after 1–2 weeks.

The Fourth Release

The fourth release involved a genetically 'crippled' AcNPV baculovirus into which a 'junk' gene had been inserted. The objective was to assess the level of expression of a foreign gene *in situ* and to determine if such levels would be suitable for a gene that might change the phenotype (characteristics) of the virus.

A genetically 'crippled' AcNPV was prepared with the gene for the bacterial enzyme beta-galactosidase substituted in place of the coding region of the polyhedrin protein gene. The bacterial enzyme gene was included as a 'junk' gene whose level of expression could be measured in infected caterpillars. Laboratory analyses established that the 'crippled' 'junk' gene virus exhibited the same host range phenotype as the marked and 'crippled' virus described above. Studies that followed on the stability of the virus demonstrated that the virus was genetically stable and did not lose the inserted gene upon successive passage through cabbage looper moth caterpillars and in tissue culture. Physically, however, the 'crippled' 'junk' gene virus exhibited the same susceptibility to degradation in soil or on leaf surfaces as the marked and 'crippled' virus described above.

Following permission to perform field tests, caterpillars infected with the 'crippled' 'junk' gene virus were placed on leaves of sugar beet plants in the field facility at Wytham in Oxfordshire, and the infections and recovery of virus monitored as a function of time. The caterpillars died from virus infection and the gene for the beta-galactosidase enzyme was expressed.

Summary and Conclusions

In summary, the results from this programme of research demonstrate that an innocuous genetic marker is a suitable tool for the identification of a genetically engineered virus released into the environment. They also showed that a genetically 'crippled' virus can be prepared that is rapidly degraded in the environment after killing its caterpillar host. Such a virus is a suitable candidate for further genetic engineering, for example for the preparation of viruses which act more quickly than the natural virus (through the incorporation of genes of insect-specific toxins, or hormones, or other genes).

Various factors were considered in order to minimise risk to the environment from the releases. These included the choice of virus, the choice of target pest, the design of the engineered viruses, the design of the field site and the procedures employed to conduct the experiment (including the final disinfection of the site). After preparing the engineered viruses, they were subjected to extensive laboratory analyses. From the data obtained we could not identify any *measurable* risk associated with the proposed experiments. This does not exclude the possiblity of an unforeseen risk, but none was identified in the laboratory studies undertaken prior to the field trials. With the results that were obtained in the laboratory analyses, we therefore applied to the appropriate authorities for permission to proceed with the field trials. The data were independently evaluated and, after obtaining the requisite licences, the field trials were instigated and the predictions of the laboratory studies were verified.

In considering the results of the studies that are reported here, it is important to emphasise that the objectives of the studies were to assess and minimise any risk that might be involved in the deliberate release into the environment of a genetically engineered virus insecticide. The study is part of a programme of scientific research, rather than the commercial development of a product. As such, the results and information derived from the studies *only apply to baculovirus insecticides*, and only to the particular virus that was used. Having said that, in order to be scientifically accurate, there is good reason to believe that other baculoviruses, similarly marked or manipulated will behave in a like manner.

Acknowledgements
I wish to thank Mr C.F. Rivers, Mr P.H. Sterling, Mrs M.E.K. Tinson, Mr T. Carty and Mr M.L. Hirst for their assistance in collecting and/or rearing the insects for the host range studies. The genetic manipulations were undertaken by Drs R.D. Possee and I. Cameron of the IoV, host range tests by Ms Cynthia Allen. In 1986 moth collections and data analyses were undertaken by Mr S.M. Eley. In 1987 similar collections were undertaken by Ms C. Doyle and Mr A.C. Forkner of the IoV. Carabid surveys and analyses were undertaken by Dr J.S. Cory. Ultraviolet light measurements in the facility were conducted by Mr H.J. Killick. The support of these and other members of the IoV staff, particularly Mr P.F. Entwistle, and Mr J.S. Robertson, are gratefully acknowledged. Dr J.I. Cooper and his staff kindly provided the plants. I am indebted to Mr R. Broadbent and Mr R.M. MacKenzie who constructed the field facility.

13 Pandora's Genes: The Trojan Horse of Bio-Engineering

ANDREW LEES

We are in the middle of a bio-revolution. Human beings have interacted with biological systems since the dawn of history, and consciously manipulated them for many thousands of years, but in the latter half of the twentieth century there has been a dramatic change. Discoveries in genetics and other areas of biology are revealing the basic processes which underlie life on earth, and the translation of these discoveries into industrial tools and products is set to transform the global economy.

The most crucial of recent advances in the life sciences are our understanding of the structure of deoxyribonucleic acid (DNA), the hereditary molecule in virtually all living organisms, and our growing ability to investigate, direct and manipulate heredity and the hereditary material. Genetic engineering originated from work on micro-organisms. The first successful genetic engineering experiments were carried out using bacteria, especially that ubiquitous organism *Escherichia coli (E. coli)*, an inhabitant of the human gut. *E. coli* has become the workhorse for genetic manipulation experiments.

We now have a whole range of biotechnologies, including protoplast fusion, gene splicing, advanced fermentation technology, genetic fingerprinting and the new reproductive techniques, which are inter-related and reinforce each other, both scientifically and economically. Although the term 'biotechnology' is a 'catch-all' phrase, in recent times we can discern a gradual change of emphasis. Established biotechnologies use older biological knowledge, which is based on the structure, functioning and development of animals and plants in their ecological settings. New biotechnologies are based on an understanding of genetics and on cellular and sub-cellular processes. This latter form of biotechnology in particular is now looming over agriculture in many ways, with, in some cases, very important environmental implications.

In this chapter, I will not embark upon a grand tour of the whole gamut of this complex subject, but will concentrate on the genetic

135

engineering of microbial pesticides. After an introduction to the area, I will focus on the work on viral pesticides being carried out in the UK at the Natural Environment Research Council's (NERC's) Institute of Virology (IoV) in Oxford and then address the problem of assessing the ecological risks associated with the environmental application of microbes. Finally I consider the economic context in which the development of genetically engineered microbial pesticides is taking place.

Microbes as Pesticides

The first experiments with microbial pesticides took place in the late 1930s using the bacterium *Bacillus thuringiensis (B. thuringiensis)*. *B. thuringiensis* is still the only microbe to have any significant commercial impact; it accounts for 90 per cent of global biopesticide sales, worth US$50 million per year (ENDS 1987). Various strains of *B. thuringiensis* are cultured on a large scale for use as microbial pesticides. Whilst these strains have been subjected to selective breeding, they have not been genetically manipulated. *B. thuringiensis*-based pesticides have been used in the USA since 1961, which is now their main market. It was anticipated in the 1960s that future sales of *B. thuringiensis* pesticide would increase. However, sales levelled and subsequently declined following the introduction of a new family of chemical pesticides – the synthetic pyrethroids.

The pendulum is now swinging back from chemical pesticides to biological pesticides because conventional agrichemicals have created environmental problems which have become an important public issue. In recognition that the pesticide market fluctuates between biological and chemical pesticides, agrichemical companies are hedging their bets and 'insuring' themselves by investing in both biological and chemical pesticide development.

In the face of increasing public concern and, in some areas, more effective regulation of chemical pesticides, from industry's point of view it is becoming exorbitantly expensive to obtain approvals for the use of new pesticide-active ingredients. It can take between ten and twelve years and cost up to US$50 million to develop a new pesticide product and get it onto the market. Many established chemical pesticides have not been properly tested by today's standards, and they are becoming less profitable as their patents expire. The development of crops tailored genetically so that they are adapted to particular chemical regimes is the ideal form of total market control, so there is an obvious incentive for the chemical industry to invade the whole field of biotechnology through 'vertical merger'.

Viral Pesticide Research

I now wish to consider the work being carried out at the IoV in Oxford under the leadership of Dr David Bishop because it illustrates some of our concerns about genetically engineered microbial pesticides. The IoV team was responsible for the first authorised deliberate release of a genetically manipulated organism in the UK, a release which took place in 1986 (see Chapter 12), and playing the leading role in the development of the public debate in the UK about the deliberate release of genetically engineered pesticides.

The baculoviruses which are the subject of investigative work at the IoV are especially interesting as pest control agents. They are thought to be inherently safe to man and other vertebrates. Some of these baculoviruses are highly pathogenic to insects, which they enter at the larval stage. They have a narrow host range, which means they are often specific to a particular insect species or a small number of closely related species. They are relatively easy to culture in large volumes. These properties give baculoviruses commercial potential as insecticides.

A feature of the particular sub-group of baculoviruses on which the IoV are working is their protein coat which renders them stable in the environment and could endow them with a long shelf-life as commercial pesticides. A critical component of the IoV experimental research strategy has been to 'cripple' these baculoviruses by removing their protein coat using genetic manipulation, thereby making the viruses less stable in the environment and more likely to be killed by conditions in the soil or by ultraviolet sunlight.

According to Bishop the purpose of removing the protein coat from the baculoviruses is to minimise their potential hazard to the environment (see Chapter 12). However, the 'crippling' of these viruses would seem to be largely motivated by 'political' rather than scientific considerations because Bishop freely admits that the procedure might be reversed upon commercialisation. This is because the virus's inability to persist in the environment reduces its effectiveness as a pest-control agent, and without its protein coat the baculovirus does not adhere to the places on the plant where it will be eaten by the larvae (caterpillars) of the target insect pest. In addition Bishop refers to the viruses which have had their protein coats removed as 'self-destructing', whereas they are in fact simply weakened – a scientific inaccuracy, and an indication of a mendacious approach to public relations (see Chapter 12).

One of the objectives of the baculovirus pesticide research project at the IoV is to make the baculoviruses more toxic and pathogenic to insects so that they are more effective insecticides and thus commercially attractive. Several ways of achieving this have been suggested including the insertion into the viral genome of an insect hormone gene, or the insertion of the toxin genes found in *B. thuringiensis* or in scorpions. Bishop estimates that a minimum of a further five to ten years of risk assessment testing is needed before such viral pesticide products could be brought to the market.

To date, all the deliberate releases of genetically engineered organisms have been, to some extent, confined. The IoV will probably be the first organisation to request permission for an unconstrained release of a genetically engineered organism in the UK. The cognoscenti, the people close to the decision-making process, see Bishop's work as laying the path for others to follow. In Bishop's own words: 'The first unconstrained release will be the bullet that the Government's Advisory Committee on Genetic Manipulation (ACGM) must bite on' (see ENDS 1988).

The deliberate release work of the IoV has important implications for the future of the genetic engineering industry because, if serious adverse environmental implications arise as a result of the IoV experiments, the public will not find these particular initiatives acceptable as a precedent for the industry. Bishop is endeavouring to make certain that no serious problems with his work will be found. But is his risk assessment work on genetically engineered baculoviral pesticides a Trojan horse for the genetic engineering industry? Is the work now being carried out at the IoV typical of the range of activities which could be subsumed under the heading 'genetic engineering of naturally occurring organisms to be released as living pesticides'? Moreover, will other scientists in the field be as scrupulous and operate with the same level of integrity as the IoV team? In such a fast-moving commercially dynamic and competitive forum, Bishop's work is not necessarily representative of all the work which will follow it, and his careful approach, his openness, his willingness to inform people, laudable as it is, may militate against public concern about this issue. Decisions about resource allocation are political. If the public is quiescent about the issue of the deliberate release of genetically engineered pesticides, will there be enough critical public awareness to ensure that it is high on the political agenda and that the regulators are resourced adequately to be effective?

Microbial Ecology and Risk Assessment

When undertaking risk assessment prior to the deliberate release of genetically engineered micro-organisms into the environment, especially of those organisms with potentially damaging effects, genetic engineers should endeavour to undertake a comprehensive risk analysis, incorporating not just the structure, function and behaviour of the organism which has been genetically engineered, but embracing the ecology of microbes in the widest sense. I am a field biologist and I know that we do not yet understand the factors controlling the abundance and distribution even of some relatively common plants – and they are easy to see!

The study of microbial ecology poses practical problems – it's difficult to creep around the landscape with a high-powered microscope! Indeed, many scientists stress the high levels of uncertainty of microbial ecology. Professor John Beringer, the Chairman of the Planned Release Sub-Committee of the ACGM [now the Intentional Introduction Sub-Committee] remarked at the Conference on the *Release of Genetically Engineered Micro-organisms* (REGEM) in Cardiff in 1988 (see Sussman *et al.* 1988): 'The information we have on survival and persistence [of bacteria] is almost nil. We do not know what factors microbes actually need to survive in the soil' (see ENDS 1988). Similarly, at the same conference Professor Gwyn Jones, the Director of the Freshwater Biological Association, has commented: 'We know very little of the fate of introduced micro-organisms and absolutely nothing of their effects on the ecosystem ... Very closely related species can have very different survival curves ... We do not understand the movement of micro-organisms in natural systems' (see ENDS 1988). Proponents of the release into the environment of genetically engineered organisms who take the attitude: 'it has worked out all right in the past, so why should we not do it in the future?' could be charged with relying on the 'poverty of historicism'.

The concern about ecological uncertainty is nowhere more pressing than in the field of microbial pesticides. An effective biopesticide agent must be, by definition, ecologically disruptive and many scientists accept that more stringent testing and regulation are required to control the toxicity or pathogenicity of the organisms involved. According to Professor Taub of the University of Washington in the USA, 'Engineering for efficiency and safety are two sides of the same coin – they can't be addressed simultaneously' (see Sussman *et al.* 1988). Detection and monitoring of micro-organisms in the environment is another key factor in microbial risk assessment. Uncertainty about how micro-organisms live and interact in the natural environment could be reduced if we could detect and track them reliably. In practice this is very difficult to do. Sometimes culture methods can be used to detect their presence: researchers can take a sample from the environment, incubate the materials on a growing-medium, and if they are lucky they may detect what they are looking for. But increasingly, work carried out by microbiologists indicates that certain culture methods do not reveal organisms which are actually present in the sample. Researchers may try and culture them but they do not *appear* to be present, even though they are there. Such cryptic organisms are viable in the natural environment, where they live and reproduce, but in the laboratory culture system they are apparently not.

How do microbiologists investigate the population dynamics of the micro-organisms which are released? They are improving their methods of detecting micro-organisms in environmental sampling, but as Professor Rita Colwell of the University of Maryland in the USA suggests:

> Although one could argue that it's possible to monitor with reasonable assurance an introduced genetically engineered microorganism during the initial stages of introduction, it is not clear that long-term monitoring will be feasible without a good knowledge base for the naturally-occurring species. We have not fully investigated the natural ecology of microbes. (Colwell 1988)

The Economic Context

Many of the scientists at the forefront of genetic engineering are aggressively defensive of their activities. This tendency has several roots – biology has been the poor relation to other natural sciences for most of this century, but molecular biology has recently endowed biologists with a new-found status and self-esteem. The self-imposed moratorium instigated by US molecular biologists in the early 1970s followed, in the USA at least, by strenuous and angry public debate has left a legacy of mutual mistrust between the scientists and the public. As with all new technologies opportunities and potential benefits are more apparent than potential risks or drawbacks. In these circumstances it is easy for genetic engineers to over-state the potential technical and social benefits of their technology.

Genetic engineering is a science and technology where there is a lot of commercial pressure. A number of multinational companies, especially those involved in the pharmaceutical and chemical industries, have become involved in genetic engineering. Economic pressure is propelling the technology into commercialisation at a rapid rate and is determining the direction of basic research. It is also leading to a high level of secrecy because many of the scientists working in genetic engineering are employed by multinational

companies. There is an international race between the governments and industries of different countries. The USA is currently
leading the way, but Japan and countries in Europe are also com-
peting as hard as they can.

The pressure to commercialise can lead to attempts to side-step
the existing regulation that provides a framework of safety. For
example, Advanced Genetic Sciences, which developed the 'ice-
minus' 'Frostban' bacterium designed to be sprayed on field crops
to give an added few degrees' protection against frost, made the
world's first authorised release of a genetically engineered microbe
in 1987. However, by then the company had already performed an
illegal open-air experiment with the microbes on the roof of their
laboratory. In another incident, a live genetically engineered
vaccine experiment took place in Argentina in 1986. Officials of
the Pan American Health Organisation (PAHO) carried the viral
vaccine from the USA to Argentina in a diplomatic bag, thus
evading the import laws of Argentina, and performed an experiment on a number of cattle without informing the appropriate US
or Argentinian authorities. This episode is particularly worrying
because of its implications for the testing of genetically engineered
organisms in the Third World, where regulations are more lax and
the public less well informed (see Wheale and McNally 1988, pp.
176–9).

Conclusions

The use of microbial pesticides could bring environmental benefits
if it replaces the use of chemical pesticides in agriculture. However,
it is not clear whether replacement would actually occur and, if so,
how significant a shift is likely, or which chemicals are most likely
to be replaced.

The problem is to weigh the relative risks of chemical pesticides
against the risks of genetically engineered pesticides. We know
alarmingly little about areas vital to environmental risk assessment
with respect to the use of naturally occurring microbes as pesticides. The future development of genetically engineered microbial
pesticides amplifies some of the existing concerns and raises new
ones.

To summarise some of the concerns that have been raised: we
know very little about microbial ecology; techniques for detecting,
monitoring and tracking microbes are imperfect. We know that
genes are transferred between free-living micro-organisms, and this
raises the possibility that the foreign genes we insert into genetically engineered microbes which are released into the environment
could be transferred to native micro-organisms, endowing them

with novel traits which could have potentially undesirable consequences for the ecosystem. Microbial pesticides must be ecologically active in order to do the job for which they are designed; this
raises concerns over the likelihood of ecological disruption, either
due to the released organism itself or as a result of gene transfer to
another strain or species of microbe. If ecological disruption were
to occur it could be serious, and whatever damage caused would
probably be irremediable.

The forecast increase in the commercial use of microbial pesticides, stimulated largely by the power of genetic engineering, could
result in large numbers of many types of organisms being released.
Together these multiple releases could have a synergistic ecological
effect which would probably be unpredictable.

International economic competition is driving this technological
revolution at a time when governments are increasingly de-
regulating, and commercial pressure could lead to significant
numbers of micro-organisms being released without proper risk
assessment or regulatory scrutiny. We need effective regulation of
the release of genetically engineered organisms into the environment. We need *independent* watchdogs. We need effective policing
of such regulations by personnel who are not bidding for jobs in
the private sector, and we need effective sanctions which can be
imposed on those who transgress the regulations. Most importantly
we need scientists whose integrity is not compromised by the piper
who calls the tune. We are worried that there will be too few
pipers, and that they will be the transnational companies who
already dominate too much of the world's economy.

14 Environmental Legislation in the USA

EDWARD LEE ROGERS

Over twenty years ago, the legal philosopher and scholar Scott Buchanan wrote that at this juncture: 'Technology is the unclarified phenomenon that poses the most important problems in jurisprudence. The fates of both science and government are epitomized in its dark oracular operations' (Buchanan 1962, p. 306). The diverse applications and powerful impacts of genetic engineering on every aspect of society render this statement equally applicable to the plight of legislatures and regulators today.

The new techniques of genetic engineering endow us with the ability to manipulate DNA so that nature can now be adapted to our preconceived notions of productivity and efficiency. Even before the advent of the new techniques of genetic engineering, humans had the power to alter nature through traditional breeding techniques, producing, for example, racehorses, vegetable and fruit varieties and pedigree dogs. These techniques were limited, however, in the range and kind of genetic combinations they could accomplish, and the speed with which changes could be achieved.

Modern genetic engineering techniques enable scientists to produce transgenic organisms with combinations of genes that could not have arisen in nature. One such example is the transgenic tobacco plants which glow in the dark as a result of the presence in their genomes of the luminosity gene of fire flies.

Genetic engineering is being used in an attempt to increase the productivity of crops and livestock. Researchers have genetically engineered crop plants to retard spoilage, and to provide fruit trees, vegetable plants and other crop plants with resistance to pesticides, herbicides or insect pests. Trees can now be genetically modified to grow more rapidly in monoculture forests. The olfactory [Latin *olfacere*, to smell] glands of north west Pacific Salmon have been genetically engineered so that the salmon no longer annually migrate from salt water to fresh water. Instead of returning to their native streams to spawn, these genetically engineered salmon live and feed in the oceans, thereby increasing their growth rate and value

143

in the market place. If, however, these salmon were to displace their migratory wild counterparts there could be a major disruption of the ecosystems of rivers in the northwestern states of the USA. Thus, although the genetic modification incorporated in these salmon is modest, their impact on the environment could be substantial.

In the USA both Congress and the regulatory agencies appear to be abdicating their collective responsibility to address the problems posed by genetic engineering. Congress has not, as yet, passed any legislation *specifically* governing the regulation of genetic engineering. The federal agencies are therefore attempting to subsume the regulation of genetic engineering under certain extant environmental statutes enacted before the advent of modern genetic engineering. In this chapter I shall describe the ways in which, as a consequence of the failure on the part of US policy-makers to introduce new legislation specifically for the environmental applications of genetic engineering, the unique environmental risks posed by this bio-revolution are being inadequately addressed by the US federal regulatory agencies.

Unique Risks

Genetically engineered organisms are designed to carry out tasks that natural organisms either do not accomplish at all, or do so only slowly or ineffectually (as judged by the human desire for ever greater, more efficient production) (Regal 1985; Simberloff 1985). They may persist in the environment, indeed, in most cases genetically engineered organisms must survive for a certain length of time in order to carry out the tasks for which they are designed. There is a danger that they may disperse and multiply, damaging the environment either directly or indirectly by harming or displacing other organisms or disrupting abiotic phenomena and cycles upon which other organisms are dependent (Vitiusek 1985). Should they become hazardous, it may not be possible to contain or recall such organisms. This problem resembles the difficulty we have in controlling or eradicating exotic (foreign) organisms which have become pests after they are either intentionally or unintentionally introduced into a new environment. Examples of such problematic introductions include the starling and the Gypsy Moth, and those micro-organisms responsible for Dutch Elm disease and the American Chestnut blight (Alexander 1985a, 1985b; Regal 1985; Simberloff 1985; Vitiusek 1985).

Predictive ecology aims to forecast environmental risks. However, it is an underdeveloped science. The environmental risks posed by genetically engineered organisms differ radically from those posed by chemicals and toxic substances. The problem of developing a predictive ecology is particularly difficult for the releases of genetically engineered micro-organisms, such as bacteria or viruses. Much controlled and contained testing would be required in an attempt to ascertain what characteristics genetically engineered micro-organisms might express in the natural environment. The costs of such research could be substantial. Were genetically engineered organisms released into the environment to cause harm – a matter that may not be detectable except over an exceedingly long period of time – there may be no remedy.

Regulatory Violation

The environmental hazards that may arise from the production and use of genetically engineered organisms and products are regulated under the broad procedural mandates of the National Environmental Policy Act (NEPA). The NEPA is a generic environmental protection statute requiring agencies to follow certain procedures when they carry out actions having impacts on the environment. This statute has been a strong catalyst in motivating the agencies to regulate genetic engineering. In the autumn of 1983 Jeremy Rifkin, Director of the Foundation on Economic Trends, successfully sued the NIH for approving a proposed experiment by the University of California without an adequate environmental assessment or environmental impact statement, as required under the NEPA. The experiment involved the release into the environment of mutant bacteria genetically engineered to remove their ice-nucleating capacity. The NIH had also failed to produce a programmatic environmental impact statement for its progamme of approving experimental deliberate releases of genetically engineered organisms, which Rifkin charged was required under the NEPA.

In May 1984 the court issued an injunction on the experiment in question and on the approval of the release of any other genetically engineered (recombinant) organisms by the NIH pending the development of the programmatic environmental impact statement (*Foundation on Economic Trends* v. *Heckler* 1984). In February 1985, on appeal by the University, the court of appeals sustained the injunction against the university's experiment (*Foundation on Economic Trends* v. *Heckler* 1985). On the NIH's appeal, the court held that there was not at that time a sufficient number of applications for deliberate releases of genetically engineered (recombinant) organisms into the environment to demonstrate a clear need for a programmatic environmental impact statement; accordingly, that part of the district court's decision was removed.

Existing Legislation

The Reagan administration attempted to remedy the confused state of genetic engineering regulation through the development of its Co-ordinated Framework for the Regulation of Biotechnology (OSTP 1986). Despite this initiative, considerable confusion has arisen over the regulation of certain categories of genetically engineered organisms. In some cases, regulatory agencies have asserted jurisdiction over certain categories of genetically engineered organisms even when it is questionable whether their existing statutory power extends that far. For example, the Toxic Substances Control Act (TSCA) authorises the Environmental Protection Agency (EPA) to regulate chemical substances which could pollute the environment. The EPA has asserted that under the TSCA it has the authority to regulate any living micro-organism developed or modified that could be a toxic substance (other than those that are pesticides, which it regulates under the Federal Insecticide, Fungicide and Rodenticide Act (FIFRA). The TSCA, however, was enacted to give the EPA regulatory authority over chemical substances, which present problems that are different in kind from those posed by genetically engineered organisms. The EPA's argument, that the statutory language giving it authority to regulate a 'new chemical substance' with a 'particular molecular identity' was intended to include genetically engineered living organisms, is tenuous (McGarity 1985).

The genetic engineering industry may choose not to challenge the EPA's asserted jurisdiction over genetic engineering under the TSCA. The industry may prefer the existing statutory scheme to a more ambitious one that the US Congress might enact were the TSCA found not to encompass genetically engineered micro-organisms. Even if the EPA were found to have the authority to regulate living organisms under the TSCA, there are other serious problems associated with the regulation of genetic engineering under this statute. For example, the language of the statute expresses Congress's clear intent that research and development activities involving small amounts of chemical substances be exempt from regulation under the TSCA. Research and development activities involving even small amounts of genetically engineered organisms, however, present unique and often serious risks. The EPA may attempt, through proposed rules, to eliminate this exemption when the research activities in question involve genetically engineered organisms. Doing so, however, requires a tortuous reading of the statutory language that may be vulnerable to legal challenge. The application of the TSCA for the regulation of genetically engineered organisms is also problematic because it places a substantial burden of proof on the EPA before the agency may request additional data that it may need to evaluate a particular organism, or before it can regulate or prohibit the production or release of a micro-organism (Mellon 1988, pp. 46–7).

Responsibility

By contrast with the EPA, other agencies have been eager to yield jurisdiction over genetically engineered organisms and products when their jurisdiction conflicts with that asserted by another federal agency. For example, the National Institutes of Health (NIH) has indicated that it will defer to the review and decision processes of other agencies that have asserted statutory jurisdiction, even if the NIH Recombinant DNA Advisory Committee (RAC) would have normally conducted a review of the genetically engineered organism or product, and when the RAC's expertise would have been of critical importance.

Even when a genetically engineered product or process appears to fall exclusively within a particular agency's regulatory jurisdiction, it may be reluctant to exercise its authority. For example, the Food and Drug Administration (FDA) has the statutory authority to approve, through a permit process, certain drugs, medical devices, biologic substances, and food and colour additives, including those produced by the various genetic engineering techniques. It has broad authority under the Food, Drug and Cosmetic Act (FDCA) to protect the public from diseases and toxins, including those arising from adulterated food and food additives. In recent years several large companies, including Calgene and Monsanto, have developed genetically engineered food plants, such as Calgene's 'super tomato' – a tomato that does not rot, which such companies now plan to make commercially available. These new food items are genetically novel and therefore their characteristics and human health consequences may be different from their natural counterparts. The FDA could assert jurisdiction over them under the FCDA on the grounds that they constitute food additives or that the genetic engineering changes involved constitute adulteration of the original food item. However, the FDA has not done so.

Filling the Regulatory Vacuum

Genetic engineering poses environmental hazards that are different in kind from those contemplated at the time the existing statutory and regulatory scheme was enacted. Existing environmental statutes that were enacted before modern genetic engineering became a reality do not adequately address these unique hazards, and the US Congress has failed to enact any legislation dealing directly with the issues that genetic engineering raises (Alexander 1985a, 1985b). The federal government has failed to fund the research necessary for the development of predictive ecology that would allow federal regulators to evaluate adequately the environmental risks posed by genetically engineered organisms. Ambitious research proposals requested by the EPA in 1985 might have led to the development of predictive ecology, but unfortunately they were never acted on (Alexander 1985a, 1985b). While the Co-ordinated Framework (OSTP 1986) has clarified certain jurisdictional questions and conflicts, it has not remedied jurisdictional confusion and overlap, for example in genetic engineering research activities (see also Wheale and McNally 1988, Chapter 8). Perhaps more disturbingly, numerous categories of genetically engineered animals, plants and microbes continue to fall outside of the US regulatory structure (McGarity 1985; Mellon 1988, pp. 48–9). At this time, therefore, given the lack of an adequate predictive ecology and the confusion over jurisdiction discussed above, the existing regulatory framework is failing to fulfill its most primary and immediate objective – to ensure the protection of the environment.

The regulatory vacuum left by US federal government is beginning to be filled by state regulations. For example, North Carolina recently passed comprehensive legislation governing the deliberate release of genetically engineered organisms into the environment (HR 748 1989). A spokesperson for the Environmental Defense Fund (EDF), a national environmental organisation that lobbied for the bill's passage, commented that, 'serious gaps and inconsistencies' within the current federal regulatory scheme necessitated action on the state level (EDF 1989).

Under the current regulatory scheme, with decision-making power resting in the hands of a small cadre of scientists, venture capitalists and government bureaucrats, most of whom have a vested interest in the growth of the genetic engineering industry, the prevailing philosophy has been that genetic engineering is simply another tool with which we can continue our practice of dominating, controlling, exploiting and manipulating nature according to our priorities and precepts of what is good and, one might add, most profitable.

This philosophy need not and should not take on the character of a moral imperative. Little research has been focused on the alternative approach of attempting to adapt our agricultural practices, our desire for perfectability and our lifestyles to nature. Preservation of nature for its own sake is itself a legitimate ethical value that must be considered alongside efficiency and productivity as part of the regulatory process (see Sagoff 1988, p. 66). An effective regulatory structure must foster debate and public participation. It should encourage alternative philosophies that would allow nature to remain largely as it is while attempting to adapt our agricultural practices and lifestyles to it (see Sagoff 1988, p. 67).

Genetic engineering endows us with a new and awesome power and, once a technology exists, it tends to be used. In my opinion, the problem is that without adequate regulation, genetic engineering will be abused for ill-conceived and irresponsible motives.

Discussion III

Andrew Lees, Friends of the Earth: I would like to ask Dr Bishop about the nature of his work for the US Department of Defence (DoD).

Dr David Bishop, IoV: When I was working in the USA I developed a candidate virus as a vaccine for Rift Valley Fever, which is a major disease of humans and animals in parts of Africa. The vaccine was developed using classical procedures, not genetic engineering. At the Institute of Virology (IoV) we are characterising this virus. Our objective is to develop it as both a human and a veterinary vaccine, although I do not think that the US Army has a cavalry any more!

I would like to respond to a point which Andrew Lees raised in his presentation. This concerns the economic and commercial pressures in the field of genetic engineering. It is extremely important that people who undertake risk assessment research, such as ourselves at the IoV, are not commercially involved but have the resources to conduct the research properly and independent of commercial interests. It is extremely important that adequate public funding be made available to do this kind of research correctly, because it would be too easy to do it incorrectly.

Ruth McNally, Bio-Information (International) Limited: Dr Bishop mentioned the IoV's plan to increase the speed with which the AcNPV baculovirus kills caterpillars through the use of genes for caterpillar hormones. Elsewhere it has been suggested that the IoV might endeavour to achieve this acceleration of the insecticidal action of the baculoviruses by using the gene for the insecticidal toxin of *Bacillus thuringiensis* (*B. thuringiensis*) or perhaps the gene for scorpion toxin.

Dr Bishop also described how, on account of the persistence of the baculovirus in the environment, it was necessary to genetically 'cripple' it by removing the genes for its polyhedrin coat protein. However, elsewhere it has been suggested that for commercial

150

applications of the virus the genes for the polyhedrin coat would be reinstated.

Dr David Bishop, IoV: At the IoV we are certainly considering the genes for *B. thuringiensis* toxin and for insect-specific scorpion toxin as candidates for insertion into baculovirus insecticides. Both of these candidate genes have been studied as potential pest-control agents. *B. thuringiensis* toxin is the preferred control agent for many insect species, particularly in Canadian forests. We are not particularly sanguine about the suitability of inserting genes from organisms other than the target insect, namely the caterpillar. We prefer to insert genes from the caterpillar itself, for example caterpillar hormone genes. If such hormone genes are expressed by the virus, we shall be very thorough in our investigations into whether they increase the host specificity of the virus.

I can see, from a commercial point of view, that a genetically engineered viral insecticide with a polyhedrin coat would be a more effective insect control agent than a 'crippled' virus because it would persist in the environment longer. But I prefer to use a 'crippled' virus that will not persist in the environment. We know that the 'crippled' virus is an efficient insecticide. We are currently investigating whether it can be delivered on to a crop in such a way as to protect the crop against a pest. It is certainly possible to put the polyhedrin coat back on to 'crippled' viruses, and this is an option that we might investigate in the future.

David Haughey: I would like to ask Dr Bishop whether animals other than insects can become infected with the baculoviruses which he is releasing into the environment, and, if they can, whether these animals sero-convert [develop antibodies to the virus] and whether these animals become carriers of the virus.

Dr David Bishop, IoV: At the IoV we have made a long-term study [some 10–15 years] of the natural course of infection between these viruses and their insect hosts in relation to other animals that prey and graze on these insects, for example birds and insectivorous [insect-eating] rodents.

Our results indicate that these viruses do not infect rodents or birds and that they do not cause sero-conversion in these animals. However, birds, insectivorous rodents and other insects can act as vectors of the virus, transporting it to other sites. For instance if a bird feeds on the carcass of a caterpillar that has been infected by the virus, it disperses the virus to a new site in its faeces. It is for this reason that the site we use for our field studies prevents the entry of birds and rodents.

I would like once more to point out that the results of our studies are only pertinent to our particular system. The system we are studying is not a model for the release of genetically engineered bacteria, which, unlike viruses, are free-living organisms. The viruses we are working with are parasites of specific insect species, and they can only replicate in these host species; otherwise they are inert. They can persist in the environment in their natural form, but not in the uncoated ('crippled') form. So I do not support the idea that our system is a model for bacteria or for any other organisms.

Dr Michael Fox, HSUS: I wish to give my reaction to the suffering of the caterpillars which are the targets of Dr Bishop's genetically engineered viral pesticide. We need to base our treatment of, and our relationship to, the environment on the fundamental principle of not causing harm, through the development of technologies and ways of living that avoid causing harm. Viral pesticides are being developed and used primarily to sustain monoculture agriculture and forestry which are not ecologically sound and are extremely vulnerable to 'pest' problems. We created the 'pest' problems. Instead of relying on genetic engineering to solve the problems we create, we need to adopt a different paradigm or way of producing food and fibre.

Harry Walters: Edward Lee Rogers argued that once a technology exists, it *tends* to be used. There is a lot of difference between this statement and the argument that once a technology exists it *has* to be used, which, in my opinion, sets up a very false technological ideal. In my long experience in the field of medicine, I have seen hundreds of technologies that we have not *had* to use, for one reason or another. A technology does not have to be used; one has the right to exercise one's judgement, including one's moral judgement, over whether or not to use a technology.

Edward Lee Rogers, Foundation on Economic Trends: What I was referring to in my presentation were those technologies that seem to be irresistible in terms of the market place. When the market demand for such technologies is great, then there is an imperative that they be used.

Anon: I should like to comment on the phenomenon of technology determining our ethical standards and give a recent example of this. In 1985 the Royal College of Obstetricians and Gynaecologists Working Party published a report in the UK entitled

'Foetal viability in clinical practice'. This report recommended that the abortion time limit be reduced from 28 weeks to 24 weeks. They chose the limit of 24 weeks because of advances in neonatal technology which enable premature babies to be kept alive from this age. They could have chosen a limit of 20 or 22 weeks but they did not do so because they wanted to allow time for a woman to have the option of an abortion, following amniocentesis and prenatal diagnostic tests. The point I am emphasising is that the state of the technology for prenatal screening and neonatal care set the ethical standard for the working party.

References III

ACGM/HSE/Note 3, *The Planned Release of Genetically Manipulated Organisms* (London: ACGM).

Alexander, M. (1985a) 'Ecological consequences: Reducing the uncertainties', *Issues in Science and Technology*, vol. 1, pp. 57–64.

Alexander, M. (1985b) 'Spread of organisms with novel genotypes', in Teich *et al.* (eds) *Biotechnology and the Environment: Risk & Regulation* (Washington DC: American Association for the Advancement of Science).

Bishop, D.H.L. (1986) 'UK release of genetically marked virus', *Nature*, vol. 323, p. 496.

Bishop, D.H.L. *et al.* (1988) 'Field trials of genetically engineered baculovirus insecticides', in M. Sussman *et al.* (eds) *The Release of Genetically Engineered Microorganisms* (London: Academic Press).

Brown, N. and Faulkner, P. (1977) 'A plaque assay for nuclear polyhedrosis viruses using a solid overlay', *Journal of General Virology*, vol. 36, pp. 361–4.

Buchanan, S.M. (1962) 'Rediscovering Natural Law', in S.M. Buchanan, *So Reason Can Rule* (1982) (New York: Farrar, Straus & Giroux, Inc.).

Bundestag Document (1987) *Report of the Commission of Inquiry on Opportunities and Risks in Genetic Engineering* (Bonn: 10/6775).

Carruthers, W.R. *et al.* (1989) 'Recovery of Pine Beauty moth (*Panolis flammea*) nuclear polyhedrosis virus from pine foliage', *Journal of Invertebrate Pathology*, vol. 2.

Colwell, R.R. (1988) 'Detection and monitoring of genetically engineered micro-organisms', in M. Sussman *et al.* (eds), *Proceedings of the First International Conference on the Release of Genetically Engineered Microorganisms* (London: Academic Press).

Dickman, S. (1989) 'Europe avoids moratorium', *Nature*, 8 June, vol. 339, p. 413.

DoE (1989) *Proposals for Additional Legislation on the Intentional Release of Genetically Manipulated Organisms: A Consultation Paper* (London: DoE).

Doyle, J. (1985) *Altered Harvest* (New York: Viking Penguin).

EDF (1989) 'Press release: Environmental Defense Fund praises landmark state action', 10 August.

ENDS (1987) 'Biological pesticides: edging into the plant protection business', *ENDS Report*, no. 151, pp. 9–12.

ENDS (1988) 'Scientists disagree over environmental hazards of genetic engineering', *ENDS Report*, no. 159, pp. 16–18.

Ellstrand, N.C. (1988) 'Pollen as a vehicle for the excape of engineered genes?', in *Planned Release of Genetically Engineered Organisms (Trends in Biotechnology/Trends in Ecology and Evolution special Publication)* (Hodgson,

J. and Sugden, A.M. eds), (pp. S30-2), Elsevier Publications Cambridge.

Entwistle, P.F. *et al.* (1983) 'Epizootiology of a nucleopolyhedrosis virus (Baculoviridae) in European Spruce Sawfly (*Gilpinia hercyniae*): Spread of disease from small epicentres in comparison with spread of baculovirus diseases on other hosts', *Journal of Applied Ecology*, vol. 20, pp. 473–87.

Entwistle, P.F. and Evans, H.F. (1985) 'Viral control' in L.I. Gilbert and G.A. Kerkut (eds), *Comprehensive Insect Physiology, Biochemistry and Pharmacology* (Oxford: Pergamon Press) vol. 12, pp. 347–412.

European Commission (1988) *Proposal for a Council Directive on the Deliberate Release to the Environment of Genetically Modified Organisms* (Brussels: COM (88) 160 final – SYN 131).

Evans, H.F. and Harrap, K.A. (1982) 'Persistence of Insect Viruses' in A.C. Minson and G.K. Darby (eds), *Virus Persistence* (Cambridge: Cambridge University Press) SGM Symposium 33, pp. 57–96.

Finney, D.J. (1971) *Probit Analysis* (Cambridge: Cambridge University Press).

Foundation on Economic Trends v. *Heckler* (1984) 587 F. Supp. 753 (DC Cir. 1984).

Foundation on Economic Trends v. *Heckler* (1985a) 756 F. 2d. 143 (DC Cir. 1985).

Gen-ethic Network (1989) 'Proposal for NGO statement on the deliberate release of genetically manipulated organisms in the environment', Conference entitled *Deliberate Release into the Environment of Genetically Engineered Organisms*, Brussels, 22 and 23 February.

GMAG Note 7 (Revised May 1979) (London: HSE), paragraph 7, p. 8. HSC (1987) *Review of the Health and Safety (Genetic Manipulation) Regulations 1978* (London: HSE).

HC Bill 14 (1989) Environmental Protection Bill (London: HMSO).

HR 748 (1989) 'An act to regulate the release and commercial use of genetically engineered organisms', *General Assembly, North Carolina*.

Hughes, K.M. (1978) 'Virus in biological control' in M.A. Brookes *et al.* (eds), *The Douglas-Fir Tussock Moth: A Synthesis* (Washington DC: US Department of Agriculture) pp. 133–6.

Hunter, F.R. *et al.* (1984) 'Viruses as pathogens for the control of insects' in J.M. Grainger and J.M. Lynch (eds), *Microbial Methods for Environmental Biotechnology* (New York: Academic Press) pp. 323–47.

Juma, C. (1989) The Gene Hunters (London: Zed Books Ltd).

Lewis, F.B. *et al.* (1981) 'Laboratory evaluations' in C.C. Doane and M.L. McManns (eds) *The Gypsy Moth: Research Toward Integrated Pest Management* (Washington DC: US Department of Agriculture) pp. 455–61.

McGarity, T.O. (1985) 'Legal and regulatory issues in biotechnology', in Teich *et al.* (eds), *Biotechnology and the Environment: Risk & Regulation* (Washington DC: American Association for the Advancement of Science).

Mellon, M. (1988) *Biotechnology and the Environment* (Washington DC: National Wildlife Federation).

Milewski, E.A. (1986) 'Report on Recent Congressional Hearing and Study Conference on Biotechnology, *Recombinant DNA Technical Bulletin*, vol. 9, no. 1, pp. 29–44.

Newmark, P. (1989) 'Danish law to be less rigid', *Nature*, vol. 339, p. 653.

OECD (1986) *Recombinant DNA Safety Considerations: Safety Considerations for Industrial, Agricultural and Environmental Applications of Organisms*

Derived by Recombinant DNA Techniques (Paris: OECD).

OSTP (1986) 'Co-ordinated Framework for the Regulation of Biotechnology', *Federal Register*, vol. 51, no. 123, pp. 23, 301–50, 26 June.

Podgewaite, J.D. (1985) 'Strategies for field use of baculoviruses' in K. Maramorosch and K.E. Sherman (eds) *Viral Insecticides for Biological Control* (New York: Academic Press) pp. 775–97.

Poole, N.J. *et al.* (1988) 'The involvement of European industry in developing regulations', in *Planned Release of Genetically Engineered Organisms (Trends in Biotechnology/Trends in Ecology and Evolution special Publication)* (Hodgson, J. and Sugden, A.M. eds), (pp. S45–7), Elsevier Publications Cambridge.

Possee, R.D. and Kelly, D.C. (1988) 'Physical maps and comparative DNA hybridisation of *Mamestra brassicae* and *Panolis flammea* nuclear polyhedrosis virus genomes', *Journal of General Virology*, vol. 69.

RCEP (1989) *The Release of Genetically Engineered Organisms to the Environment* (London: HMSO).

Regal, P.J. (1985) 'The ecology of evolution: Implications of the individualistic paradigm', in H.O. Halvorson *et al.* (eds), *Engineered Organisms in the Environment: Scientific Issues* (Washington DC: American Society for Microbiology).

Regal, P.J. *et al.* (1989) *Basic Research Needs in Microbial Ecology for the Era of Genetic Engineering* (Santa Barbara, CA: FMN Publishing).

Rose, C. (1989) *Blueprint for a Green Europe: An Environmental Agenda for the 1989 European Elections* (London: Media Natura).

Sagoff, M. (1988) *The Economy of the Earth* (Cambridge and New York: Cambridge University Press).

Schmid, G. (1989) 'Report on the proposal for a Council directive on the deliberate release to the environment of genetically modified organisms (COM (88) 160 final – Doc. C 2-73/88 – SYN 131) *European Parliament Committee on the Environment, Public Health and Consumer Protection* (Brussels: DOC-EN\PR\65098.TO PE 128.472/fin, 28 April).

Simberloff, D. (1985) 'Predicting ecological effects of novel entities: Evidence from higher organisms', in H.O. Halvorson *et al.* (eds), *Engineered Organisms in the Environment: Scientific Issues* (Washington DC: American Society for Microbiology).

Simonsen, L. and Levin, B.R. (1988) 'Evaluating the risk of releasing genetically engineered organisms', in *Planned Release of Genetically Engineered Organisms (Trends in Biotechnology/Trends in Ecology and Evolution special Publication)* (Hodgson, J. and Sugden, A.M. eds), (pp. S27–30), Elsevier Publications Cambridge.

Sussman, M. *et al.* (eds) (1988) *Proceedings of the First International Conference on the Release of Genetically Engineered Micro-Organisms* (London: Academic Press).

UK Genetics Forum (1989) *Response to the RCEP report* (1989) (London: UKGF).

Vitiusek, P.M. (1985) 'Plant and animal invasions: Can they alter ecosystem processes?' in H.O. Halvorson *et al.* (eds), *Engineered Organisms in the Environment: Scientific Issues* (Washington DC: American Society for Microbiology).

WHO/FAO (1973) Expert Group (GTRES): *The Use of Viruses for the Control of Insect Pests*, WHO Technical Report Series 531.

Wheale, P.R. (1986) 'Food and agriculture', in C. Boyle *et al.*, *People, Science and Technology* (pp. 77–101) (Hemel Hempstead: Wheatsheaf Books; New Jersey: Barnes & Noble).

Wheale, P.R. and McNally, R.M. (1988) *Genetic Engineering; Catastrophe or Utopia?* (Hemel Hempstead: Wheatsheaf; New York: St Martin's Press).

Wheale P.R. and McNally, R.M. (1990) *'Quis custodiet ipsos custodiens?*: Regulating the environmental release of genetically engineered organisms', *Rivista Giuridica Dell'Ambiente,* (in press).

Wigley, P.J. (1980) 'Diagnosis of virus infections-staining of insect inclusion body viruses' in J. Kalmakoff and J.F. Longworth (eds) *Microbial Control of Insect Pests* (New Zealand; New Zealand Department of Scientific and Industrial Research Bulletin) vol. 228, pp. 35–9.

Williamson, M. (1988) 'Potential effects of recombinant DNA organisms on ecosystems and their components', in *Planned Release of Genetically Engineered Organisms (Trends in Biotechnology/Trends in Ecology and Evolution special Publication)* (Hodgson, J. and Sugden, A.M. eds), (pp. S32–5), Elsevier Publications Cambridge.

Part IV
Philosophical, Theological and Ethical Responses to the Bio-Revolution

15 Introduction IV

Dr PETER WHEALE and RUTH McNALLY

Genetic engineering endows us with a new order of power over the natural world. But do we have the right to manipulate and control the natural world in whatever way we choose or should our activities be subject to certain moral constraints? And if so, what reasoning should guide us in our decisions concerning the restrictions we place on the practice of genetic engineering?

In this part of the book Alan Holland, a Lecturer in Philosophy at Lancaster University and a member of the Society for Applied Philosophy; the Reverend Dr Andrew Linzey, Chaplain and Director of Studies at the Centre for the Study of Theology, Essex University; Richard Ryder, Chairman of the Liberal Democratic Party Animal Protection Group and a former Chairman of the RSPCA Council; and Peter Roberts, the founder of Compassion in World Farming and an Honorary Director of the Athene Trust, consider these questions from ethical, philosophical and theological points of view.

Do Sentients have Moral Rights?

The Cartesian notion that non-human animals do not feel pain has long since been discredited and it is now recognised that all vertebrate species with central nervous systems feel pain. The term 'sentient' refers to all animals which have the capacity to have conscious experiences such as pain or pleasure.

In *Animal Liberation* (1977) Peter Singer argues that all sentient animals should be regarded as morally equal. He considers that a philosophy which allows rights to humans but denies those rights to other sentient animals is a moral fallacy analogous to sexism or racism, a fallacy called 'speciesism'. Speciesism insulates animal experimenters from the ethical consequences of their actions because it reinforces the perceived divide between us and other animals. In recent times 'speciesism' has been challenged by studies in neurobiology which have furnished evidence that humans share many cognitive and perceptual attributes with other members of the animal kingdom.

161

There is a distinction between having regard for the welfare of animals and believing that animals have inviolable moral rights. As a utilitarian, Singer (1977) prefers to avoid speaking of the moral rights of animals, at least in so far as these are construed as claims which may sometimes over-ride purely utilitarian considerations. According to this philosophical position, costs in terms of animal suffering, and benefits in terms of human gain can be compared, and in certain circumstances animal suffering can be justified.

In *The Case for Animal Rights* (1983) Tom Regan argues that non-human animals do have inviolable moral rights, the most important of which is the right to life. In Chapter 18 Richard Ryder, arguing on similar moral grounds to Regan, contends that it is never justifiable to trade off the suffering of one sentient animal against the pleasure of another. He calls his position 'sentientism', a morality which holds it wrong to cause pain or distress to any sentient animal unless it is with their agreement or unless it will bring unquestionable benefit to that same individual sentient if he or she is unable to give informed consent. According to the philosophy of sentientism, the transgenic manipulation of animals is morally wrong only if it causes suffering to the animals involved (see also 'Discussion IV'). The reader may be interested to compare Ryder's philosophical stance with that expressed by Dr Michael Fox in Chapter 5 above. In Fox's view, transgenic manipulation is morally wrong because it violates the genetic integrity or *telos* of organisms or species, and its use can only be justified in very special circumstances, for example if it were the only way to produce a life-saving drug.

The scientific demonstration of the similarity between humans and other sentients could be invoked to justify calling a halt to experiments which severely impair the welfare of animals. On the other hand, if in the light of this new knowledge we continue to treat non-human animals in an inhumane way, then, as Ryder argues, logically there will be even less reason than previously why we should not similarly exploit our fellow human beings.

The Reverend Dr Andrew Linzey, in Chapter 17, perceives genetic engineering to be a means of subjugating the nature of animals so that they become human property. Linzey, as a theologian, expresses the view that human beings cannot justify claims of absolute ownership rights over animals for the simple reason that God alone owns creation. He quotes from George Orwell's *Animal Farm* (1961) to illustrate the similarity between the arguments which seek to justify the enslavement of humans and those which are used to justify the oppression of non-human animals. As Linzey reminds us, Aristotle held that the conquest of human 'natural slaves' was right and that non-human animals belonged to, and existed for, their human masters.

In Chapter 16 Alan Holland argues that the ethical issues involved in genetic manipulation go beyond the question of suffering, and involve questions of harm and freedom. He maintains that restrictions should be placed upon those practices which involve taking a form of life with a given level of capacity to exercise options and significantly reducing that capacity. This would place constraints, for example, on the genetically engineered 'crippling' of organisms to make them safer to use in medicine, agriculture, manufacturing and resource-recovery.

In Chapter 19 Peter Roberts argues that genetic engineering has the capacity both to harm animals and, through the creation of laboratory (*in vitro*) meat factories, to liberate farm animals. He believes that the exploitation of animals for food cannot be justified now that we are able to feed the world's population with a wholesome, nutritious and varied diet using only a fraction of the present cultivable land. However, although he considers it to be theoretically possible he is doubtful that genetic engineering will be used to facilitate such an animal liberation.

The 'Land Ethic'

The ecologist, Aldo Leopold, was one of the first exponents of the view that moral rights should be extended to the non-human world including plants, animals and natural habitats (Leopold 1949). Holland adopts and develops Leopold's concept of the 'land ethic', one of the most influential ideas in the history of ecology. Under this 'holistic' ethical system, our interactions with the non-human world are considered to be morally right only when they tend to preserve the integrity, stability and beauty of the biotic community. The 'land ethic' would dictate, for example, that the earth's genetic resources be treated as a valuable non-renewable resource and therefore preserved as part of the world's genetic heritage.

Genetic engineering involves using living things as instruments which we exploit for our own devices. Holland points out that there is a distinction between using another animal's ends for one's own purposes and disregarding that other animal's ends entirely. He does not, however, reject genetic engineering on the grounds merely of 'instrumentalism' because he reasons that treating a living thing as an instrument is not *necessarily* incompatible with showing it respect. Holland identifies a number of constraints to which the practice of genetic engineering should be subjected if it is not to violate the biotic community which we share with other life-forms. These constraints include the need to curtail applications which involve inflicting significant suffering or harm upon other living things or significantly reducing their capabilities; the need to guard against the loss of genetic diversity; and the need to abandon utopian visions of making the world completely safe for human life.

Future-Oriented Ethics

In *The Imperative of Responsibility* the philosopher, Hans Jonas, argues that we urgently need a future-oriented ethics which recognises our responsibilities to the posterity of humankind and the environment (Jonas 1984). According to this view, there is a need to enshrine future-oriented ethics into our present-day regulations to ensure that we do not deplete, pollute and pervert the stock of biological resources for future generations. As Roberts points out, current regulations do not address the ethical questions raised by genetic engineering.

There is a danger that genetic engineering may accelerate the process of the depletion of the gene pool. Genetically engineered 'super animals' and 'super crops' threaten to colonise and destroy the few remaining wildernesses. Genetically engineered 'super plants' will be used to replace indigenous plant varieties, whose unique genetic complement, the product of centuries of evolution and, in some cases, careful cultivation, may be lost forever. Genetically engineered pesticides could either deliberately or unexpectedly eradicate species. The genetic 'crippling' of organisms selected and engineered for their utility could result in the colonisation of farms and factories, rivers, seas and mines with organisms whose vital functions are tailored so that they can only undertake the tasks they have been designed to perform.

Our Common Future (Brundtland 1987), the report of the United Nations (UN) World Commission on Environment and Development chaired by Gro Harlen Brundtland, the former Norwegian prime minister, recommends that the concept of 'sustainable development' be employed to guide the development of a future-oriented environmental ethics which will ensure that future generations inherit an uncompromised environment.

'Sustainable development' (with its corollary of 'sustainable income') requires that each generation pass on to the next generation an undiminished aggregate of capital assets, including certain environmental assets which should be inviolable, such as our stock of biological diversity. This will only be achieved by giving the protection of the environment a much higher weighting in policymaking than has been the case to date. But can an ethics which is future-oriented be developed and implemented within existing liberal representative democracies? Many believe it can. It has been suggested, for example, that environmental factors could be weighted so as to bias policy decisions against irreversible choices, in favour of offering special protection to those who are especially vulnerable to our actions and choices, in favour of sustainable benefits, and against causing harm as distinct from merely forgoing benefits (Goodin 1983).

16 The Biotic Community: A Philosophical Critique of Genetic Engineering

ALAN HOLLAND

Existentialist philosophers who talk about the absurdity of the human condition highlight a fracture – a division or dualism – in our human perception. On the one side we are active, involved beings convinced of the meaning and purpose of our individual schemes and projects, which we prosecute with the utmost intensity and commitment. On the other side we are capable of commanding an 'external' – a more detached – viewpoint, in the light of which these very same projects can appear trifling or futile. That is our plight as human beings, our absurd condition, and that is how we have evolved to be.

The extent to which we are in two minds about genetic engineering is also due, I suggest, to some such *unavoidable* fracture in our human perception. For at their best, and in principle, the techniques of genetic engineering may be used to serve quite worthy purposes, mainly, so far, to produce medicines and to enhance and protect farm animals and crops from diseases. These purposes are centred firmly on perceived human needs, and it cannot be a matter of reproach that humans should seek to alleviate or ameliorate their condition. This, at any rate, is how things seem from one perspective – the anthropocentric perspective.

At the same time, our powers of reflection have given us the capacity to conceive another perspective, an external and holistic one, in the light of which these erstwhile worthy purposes become open to critical review. The current environmental crisis has indeed forced us to take this more holistic perspective very seriously, and has prompted some harsh self-assessment. For example, some critics have gone so far as to liken the human presence on the planet to that of a disease. As is argued in the introduction to the reputable *Gaia Atlas of Planet Management*: 'In certain senses, humanity is becoming a super-malignancy on the face of the planet' (Myers 1985, p. 20). The holistic perspective from which this sort of judgement becomes possible is essentially a biocentric one.

It would be unwise, even if it were possible, wholly to abandon, or to cleave to, either the anthropocentric or the biocentric perspectives; to do so would be to frustrate part of what we are. The risk in attempting to embrace a wholly biocentric point of view is that we may encourage misanthropic or even fascist sentiment. The results of clinging too steadfastly to the anthropocentric point of view are only too visible in today's environment with its pollution problems. However, given that it is incumbent upon us that we live with both perspectives, how can we best do this?

The Biotic Community

One suggestion that I want to develop in this chapter involves taking seriously the exhortation of the US environmentalist Aldo Leopold that we should look upon what he calls the 'land' not as a commodity which we own but as a 'community to which we belong' (Leopold 1949, p. viii). Leopold defines 'land' collectively as 'soils, waters, plants, and animals' (Leopold 1949, p. 204). One way, it seems to me, in which we might accommodate our two perspectives, tempering or modifying both but abandoning neither, is to perceive human beings as belonging to a 'community' in Leopold's sense. For a community is a framework within which it is possible to pursue one's own concerns while at the same time heeding the implications of those pursuits for the community at large. Conversely, it is because we are beings with these two different perspectives that we are capable of forming and belonging to communities.

The term 'community' has vital resonances lacking in the more neutral alternative term 'ecosystem'. To say that we belong to a 'system' is simply a way of drawing attention to our interdependency with other parts of the system; it contains no hint of a situation giving rise to responsibilities. Why should we not 'play the system'? Moreover we do not want a term which suggests the submerging and losing of the human interest in that of some larger whole, for that would mean abandoning any critical stance along with all our other interests.

The bearing of Leopold's (1949) exhortation on the problems that we have with genetic engineering is to suggest that we should think of there being a general requirement *that it be conducted in a manner compatible with the continuing existence of the biosphere viewed as a community.* Such a requirement lacks precision and is susceptible to being interpreted in different ways, but in very general terms it does, I believe, provide a framework for our thinking. It should not be confused with the bland recommendation to live in harmony with nature. For a community is, typically, neither a static nor an entirely harmonious affair. It may be, for example, that no particular state of the biosphere is sacrosanct and that no life-form can expect to be immune from the hostile attentions of some other life-form.

For the remainder of this chapter, and using the notion of a biotic community as a framework, I shall consider whether there is a basis for rejecting genetic engineering in principle, and what constraints genetic engineering should be subject to.

Is There a Philosophical Basis for Rejecting Genetic Engineering?

One sincerely held religious view of genetic engineering is that it is, quite simply, blasphemy. The sentiment here is that virtually any application of these techniques steps outside of the range of practices that is appropriate to our place in the biotic community. But we need to look for some secular counterpart to the objection of blasphemy, I think, if we are to find an objection which is capable of commanding widespread assent. One such secular counterpart is to be found in the view that when we genetically engineer life-forms we go against, if not God, then nature.

Before genetic engineering emerged, the anthropologist Carleton S. Coon was able to write engagingly of the 'conversion of the sheep into a walking and bleating wool-factory', and to describe it as 'one of our ancestors' most dramatic perversions of Nature' (Coon 1962, p. 154). Others pronounce domestic animals in general to be 'unnatural'. Such charges surely attach with redoubled force to the products of modern genetic engineering. Several points here call for comment. One is that the charge against genetic engineering, that both its products and its processes are unnatural, is little more than a trivial truism. No one is denying that both product and process are unnatural in the sense that they are artificial, or human-made. The question is whether there is anything objectionable about that. Moreover if this objection is levelled against modern genetic engineering, then it has to be levelled also, as Coon (1962) implies, against conventional practices of domestication and cultivation. The question then becomes whether these practices too have to be given up before humans can be judged to be behaving in a way compatible with membership of the biotic community. The chief reason for doubting that we have to give up conventional practices of domestication and cultivation on the basis that these are perverse and unnatural practices is that they pre-date science and have their origins in the earliest human societies.

In truth, a genetic engineer with a lawyer's nose for precedent can find precedent enough in our previous practices. What does the genetic engineer do but seek to render a plant permanently toxic to certain insects – a condition which systemic insecticides already achieve for a limited time? What does the genetic engineer do but bring about at an earlier stage in a plant's development the kinds of results which are already brought about at a later stage by grafting? And when Charles Darwin writes, of a certain Mr Wicking, that he took just 'thirteen years to put a clean white head on an almond tumbler's [a variety of pigeon] body' (Darwin 1899, p. 183), what does the genetic engineer do but change the thirteen years that Darwin mentions to thirteen months?

Some critics of genetic engineering consider that it is an invasion of biological integrity, both at the level of the individual and at the level of the species. However, according to evolutionary theory, there is no such thing as biological integrity. There are only more, or less, stable life-forms. 'The tangible small-scale evidence of inherited change engineered by breeders and farmers through artificial selection among naturally-occurring variants were proclaimed by Charles Darwin (1859) to be evidence for the theory of evolution by natural selection' (Wheale and McNally 1988, p. 19). If one accepts Darwin's theory on the origin of species it becomes very difficult to argue that natural kinds and individuals have an integrity which artificial kinds and individuals lack. Nearly all of the distinctions which Darwin observes between natural and artificial forms of life – and his discussion of these distinctions remains classic – reduce to a difference of degree rather than of kind.

The one exception is Darwin's observation that the artificially selected races are modified 'not for their own benefit, but for that of man' (Darwin 1899, p. 4), but this distinction is a questionable basis upon which to argue that the integrity of artificial races is compromised. It would not be easy to show that natural selection is more considerate of its products than is artificial selection. (Is it better to be a fox or a spaniel?) It does not follow that a modification for our benefit may not also benefit the modified creature.

A clear consequence of the view that we form a community with other life-forms is that we should show them respect. What Darwin's distinction *does* remind us of is that it is of the essence of domestic animals – indeed, of the artificially selected races in general, both animals and plants – that they are viewed by us as *instruments*. The question then is whether this is compatible with showing *respect* for those life-forms. And if it is not, then we have here what seems to me the *best* case against genetic engineering in principle. For genetic engineering essentially involves using living things as instruments. Moreover, it reduces them to instruments of a particular kind – mechanical ones. It does not simply view living things as instruments; it views them as *mechanisms*. If using a life-form as an instrument is not compatible with showing it respect, then it follows that genetic engineering cannot be assimilated within the framework of the biotic community.

The matter turns on what constitutes a failure to show respect. The best articulation of this concept is that offered by the eighteenth-century German philosopher Immanuel Kant. He urged that we should so act as to treat others as ends, never as means only. He was speaking of human beings, but the point can be transferred without too much difficulty to other living things, as an admonition to respect their ends. For anything that is capable of flourishing can be construed as having that as its end. This notion seems to rule out using a life-form in any way which seriously frustrates its ends. But note that Kant argues, 'never as means *only*' (Kant 1909, p. 47); therein lies the weakness in the argument. For it means that, in Kant's view at least, treating another life-form as an instrument is *not* incompatible with showing it respect. There is a distinction between using another creature's ends as your own – which is acceptable – and disregarding that other creature's ends entirely – which is not. A problem, however, which Kant's notion does not seem to address, and which I touch on below, comes when the genetic engineer starts to redesign those ends.

I conclude that some of the objections aimed at rejecting the genetic engineering enterprise in principle seem lacking in substance and none entirely carries conviction. One objection, however, which is far from lacking in substance has not yet been mentioned. This objection views as unacceptable any applications of genetic engineering which make, or are likely to make, sentient creatures suffer. Clearly this is not a basis for objecting to genetic engineering in principle but, rather, suggests a constraint which any acceptable form of genetic engineering should satisfy.

What Constraints Should Genetic Engineering be Subjected to?

Genetic engineering puts a great deal of power in the hands of those who have control of it. As power in democratic societies needs to be divided, it follows that genetic engineering should be subject to some form of democratic regulation which would constrain the power of the genetic engineering enterprise. The question is: what should be the form of the regulations?

Of the genetic engineering experiments so far attempted which involve sentient creatures, some are directed at relatively peripheral animal functions, or at least can be reasonably predicted to have only a minimal effect on the resultant animals' welfare. A case in point, perhaps, is the secretion of blood-clotting Factor IX in the milk of sheep [see Chapter 3]. A number of other valuable substances may eventually be obtained in this way, and one particular development envisaged is to use these techniques to render cow's milk palatable to the populations of certain Third World countries who at present do not find it so.

Other experiments, however, involve central animal functions and can reasonably be predicted to have a considerable effect on the animals' welfare. An example of this kind of experiment is the introduction of a foreign growth hormone gene so as to radically alter an animal's pattern of growth. It would need very good reasons indeed to justify this sort of development and, so far as I can see, no such good reasons exist. But the particular point that I wish to bring out here is that, in formulating the constraints that should apply, we should look beyond the question of suffering. We should consider also the question of harm, and the question of freedom.

First, the scope of concern must surely be widened to include harm simply because an animal can be injured, or diseased, or rendered incapable of reproducing without necessarily suffering. Yet all these conditions constitute an animal's being harmed and in normal circumstances should not knowingly be brought about. Notice what a very powerful constraint that would be. It would extend concern even as far as the genetically engineered virus [see Chapter 12]. 'Crippling' a virus is, without doubt, harming it, and there is surely something disturbing in the idea of deliberately creating 'crippled' life-forms, that is, creating life-forms which are permanently harmed.

Secondly, suppose that animals were to be genetically engineered to be tolerant of their new conditions. Suppose, for example, that compensating structures and patterns of behaviour are introduced so that what would normally count as an arthritic and overweight pig is in no way impeded by these conditions. Suppose that such a genetically engineered animal neither suffers nor is harmed. What *then* would be the grounds for objection? There is much talk these days of intensively reared animals being 'bred to withstand their conditions', and therefore this question concerns conventionally bred animals as much as it concerns genetically engineered ones.

Freedom can be defined as the capacity to exercise options. It seems to me that we ought to be very concerned by any developments which would diminish the general level of freedom of sentient animals. The genetic engineer is unlikely to deal with the problem of the unhappy pig by increasing the animal's sophistication to the level where it is capable of being philosophical about its condition and learns not to mind. Rather, the capacities of the animal would be reduced to the state where it was, from a sentient point of view, more vegetable than animal. I suggest that in general, and within the framework of a biotic community, there should be an objection to any practice which involves taking a form of life with a given level of capacity to exercise options and reducing that capacity significantly.

It is sometimes suggested that opposition to genetic engineering is based upon a mixture of conservatism and sentimentality; perhaps it is, sometimes. Natural life-forms function like old, familiar friends and there is a certain harmless conservatism about the desire to see these retained and not tampered with. Humans, after all, have needs other than material ones, and one of these is the need to retain a sense of identity given, partly, by familiar things.

The irony is that the charge of conservatism can with more justice be reversed. I argue that a deeper and more dubious conservatism underlies the utopian vision of a custom-designed environment, created using unfettered genetic engineering technology – an environment in which surprises are reduced to a minimum (accidents excepted) and in which nothing ultimately remains which is untouched by human hand. What makes such conservatism more dubious is that, yielding to it, we risk sacrificing the refreshment of spirit and challenge which can only be enjoyed through contact with a natural world which is 'other', that is, given and not made.

Turning to the charge of sentimentality, we must distinguish between this and sentiment, or feeling. There is nothing the matter with sentiment. It is as absurd to complain, for example, that opposition to animal experiments is based upon sentiment, as it is to complain of people getting married for reasons of sentiment. Sentimentality, on the other hand, is a vice, a matter of allowing oneself to believe that the world is a more comfortable and cosy place than it is or can be made to be. The idea that we might create, through genetic engineering, a world free of all threats to human existence constitutes just such a belief. It is unreasonable to think that such a state of affairs is attainable; and the attempt to attain it, presumably by destroying or modifying all other lifeforms that could conceivably constitute a threat, would be at too great a cost to the biotic community. My point here is that, for the sake of the community as well as of ourselves, we have to give up the sentimental ideal of total security for the human race and resist the determined application of genetic engineering to that purpose.

A further source of concern and possible constraint centres upon the cumulative effect of genetic engineering on genetic diversity. There are at least two ways in which genetic engineering might contribute to the loss of genetic diversity throughout the biotic community as a whole. One is through the extensive use of cloning, where space which could have been occupied by a genetically distinct individual is occupied by an individual who is simply a genetic copy of another. The other way in which genetic diversity might be lost arises from the fact that genetic engineering will inevitably reflect human priorities, and human priorities are inevitably parochial. Life-forms which are in any way serviceable to humans will be developed in those respects in which they are serviceable and, in general, genes which express such properties will be favoured at the expense of those which do not.

Why exactly is loss of genetic diversity undesirable? Perhaps the chief reason is that genetic diversity renders a life-form or group of life-forms more resilient – better able to bear vicissitudes, whether in the form of widespread natural disasters within the biosphere or cometary showers from without. A reduction in diversity would also be against the general trend of evolution to date, which has been to elaborate and diversify the biota. This is not to suggest that there is anything inherently sacrosanct about the trend of evolution to date, but only to point to the apparent folly of reversing the trend which has made the biotic community possible.

Finally, I will consider briefly a facet of genetic engineering which is normally looked upon as one of its chief virtues – its potential use in agriculture to protect our crops and animals from pests and diseases. Genetic engineering promises us pesticide products for agriculture which are neat, clean and precisely targetted in some cases to individual species. This is in contrast to the products of the chemical industry; chemical insecticides, for example, are notorious for killing many more living things than is necessary for the purpose in hand. But consider the reverse side of this contrast between chemical and biological pesticides. Picture, if you will, the many forms of life as interwoven horizontal threads. Species-specific pesticides would be capable of removing entire threads from the fabric of the biotic community with unpredictable consequences. Unlike chemical substances, genetically engineered organisms have the capacity to mutate, migrate and multiply. It is almost impossible to guarantee that genetically engineered organisms released into the environment will remain localised. They can be genetically marked [see Chapter 12] but this is a method of recognising them when found; it is not a method of finding them. In the case of micro-organisms, they would be extremely difficult to trace.

The points discussed above suggest to me that the pest control or disease control products of genetic engineering which are targeted against specific species are among those which need to be looked at with special care. It seems paradoxical that whilst striving to save some species from extinction we simultaneously target populations of other species for destruction using genetic engineering.

Conclusions

In this chapter I have suggested that both enthusiasm for, and doubts about, genetic engineering arise from genuine human concerns. We should be prepared to accommodate a point of view in which we see ourselves as part of a 'community' with other life-forms and the structures that support them – what Aldo Leopold called the 'land' – and act accordingly. I have attempted to put genetic engineering into an evolutionary perspective from which I conclude that there is no unique argument for rejecting genetic engineering in principle. However, I have identified a number of constraints to which the practice of genetic engineering should be subjected if it is not to violate our 'community', which we share with other life-forms, and the structures which support it.

I do not know the 'right' way forward in regulating and controlling genetic engineering, and do not apologise for that because there is no such thing. There are only more, or less, sensible ways. I only hope that through dialogue between all concerned groups we shall manage to find one of the more sensible ways forward.

17 Human and Animal Slavery: A Theological Critique of Genetic Engineering

The Reverend Dr ANDREW LINZEY

'Animal Farm' and Slavery

Imagine a place called Manor Farm. The farmer, Mr Jones, has retired for the night. Quite an ordinary farm of its type in almost every way, with a wide variety of animals: cart-horses, cattle, sheep, hens, doves, pigs, pigeons, dogs, a donkey and a goat. The only difference with this farm is that the animals can talk to one another. And in the dead of the night when the farmer is sound asleep, the Old Major, a prize Middle White Boar, addresses a secret meeting in the barn. He begins:

> Now, comrades, what is the nature of this life of ours? Let us face it: our lives are miserable, laborious, and short. We are born, we are given just so much food as will keep the breath in our bodies, and those of us who are capable of it are forced to work to the last atom of our strength; and the very instant that our usefulness has come to an end we are slaughtered with hideous cruelty. No animal in England knows the meaning of happiness or leisure after he is a year old. The life of an animal is misery and slavery: that is the plain truth.

The Old Major continues his oration with increasing passion:

> But is this simply part of the order of nature? Is it because this land of ours is so poor that it cannot afford a decent life to those who dwell upon it? No, comrades, a thousand times no! The soil of England is fertile, its climate is good, it is capable of affording food in abundance to an enormously greater number of animals than now inhabit it ... Why then do we continue in this miserable condition? Because nearly the whole of the produce of our labour is stolen from us by human beings. There, comrades, is

175

the answer to all our problems. It is summed up in a single word – Man. Remove Man from the scene, and the root cause of hunger and overwork is abolished forever.

Man is the only creature that consumes without producing. He does not give milk, he does not lay eggs, he is too weak to pull the plough, he cannot run fast enough to catch rabbits. Yet he is lord of all the animals. He sets them to work, he gives back to them a bare minimum that will prevent them from starving and the rest he keeps for himself ... and yet there is not one of us that owns more than his bare skin.

Finally the oration reaches its crescendo to gladden the animal hearts that hear it:

What then must we do? Why, work night and day, body and soul, for the overthrow of the human race! That is my message to you comrades: Rebellion! I do not know when that Rebellion will come, it might be in a week or in a hundred years, but I know, as surely as I see this straw beneath my feet, that sooner or later justice will be done. Fix your eyes on that, comrades, throughout the short remainder of your lives! And above all, pass on this message of mine to those who come after you, so that future generations shall carry on the struggle until it is victorious (Orwell 1961, pp. 5–10).

By now, of course, you will have guessed the location of Manor Farm – in the *Animal Farm* of George Orwell's imagination. We all know that Orwell intended his book not as a satire on the oppression of pigs and horses but on the oppression of working class humans by their indolent and unproductive bosses. Nevertheless it could not have escaped Orwell's attention, as it may not have escaped ours, that there is indeed a great similarity between the arguments used (so brilliantly summarised and rebutted by the Old Major) for the justifying of oppression of humans and animals alike.

If we see this similarity we shall also have grasped something historically quite significant. For the two arguments, or rather assumptions, alluded to in the rousing polemic of the Old Major – namely that one kind of creature belongs to another and exists to serve the other – have not been confined to the animal sphere. You may recall that it was Aristotle who held that animals were made for human use: 'If then nature makes nothing without some end in view', he argues, 'nothing to no purpose, it must be that nature has made all of them [animals and plants] for the sake of man'

(Aristotle 1985, p. 79). Notice here that Aristotle is not claiming that we may sometimes make use of animals when necessity demands it, rather he is asserting that it is in accordance with nature, indeed it is *by nature*, that animals are humans' slaves. If we ask how Aristotle knows that animals are by nature slaves the answer seems to be that if they were not they would 'refuse', but since they do not, it obviously follows that it is natural to enslave them. It is crucial to appreciate, however, that this ingenious argument does not stand alone in Aristotle's *The Politics*. When Aristotle comes to consider the right ordering of society, based in turn on the pattern of nature, he uses the example of animal slaves to underline and justify the existence of human slaves as well:

> Therefore whenever there is the same wide discrepancy between human beings as there is between soul and body or between man and beast, then those whose condition is such that their function is the use of their bodies and nothing better can be expected of them, those, I say, are slaves by nature (Aristotle 1985, pp. 68–9).

In a notorious section, Aristotle describes human slaves as 'tools' and none other than pieces of property. '[A] slave is not only his master's slave but belongs to him *tout court*, while the master is his slave's master but does not belong to him' (Aristotle 1985, p. 65). In short: Aristotle does not demur from using the same two arguments, namely that one creature belongs to another and one kind of creature exists to serve the other – to justify both animal and human slavery. As for women, incidentally, they appear to stand somewhere in between, possessing some soul, that is reason, but not as much as men, and having a kind of half status depending upon their rationality (Aristotle 1985, pp. 67–8).

The reason why it is worth considering Aristotle is because he represents what we may call the 'belong to and exist for' element within the Western tradition which Christianity in particular has taken over and developed to the detriment of slaves and women as well as animals. In Aquinas, for example, writing a few centuries after Aristotle, we find this same argument repeating itself: 'There is no sin in using a thing for the purpose for which it is', argues Aquinas. And again:

> Dumb animals and plants are devoid of the life of reason whereby to set themselves in motion; they are moved, as it were, by another, by a kind of natural impulse, a sign of which is that they are naturally enslaved and accommodated to the uses of others (Aquinas 1918, p. 477).

Also with women, though to a lesser degree, we may observe a similar logic. Men, not men and women, are made in the image of God and thus only males possess full rationality. Women are halfway between men and the beasts. '[I]n a secondary sense the image of God is found in man, and not in woman,' argues Aquinas, 'for man is the beginning and end of woman' (Aquinas 1922, p. 289). We see an echo of Aristotle in these words.

The simple point I want to make is this: the debate about slavery, human or animal, is not over. Let us take human slavery first. Most of us think that the battle about human slavery was fought and won more than 100 years ago. If that is what we think, we are simply mistaken. The Anti-Slavery Society exists to combat slavery which continues to exist in many parts of the world, albeit under different guises and in different forms. Based in Brixton, South London, it currently campaigns against chattel slavery, debt bondage, serfdom, child exploitation and servile forms of marriage. But if we examine the issue of slavery and the slave trade of less recent history and go back to previous centuries we will find intelligent, respectable and conscientious Christians supporting almost without question the trade in slaves as inseparable from Christian civilisation and human progress. The argument is not an exact repeat of Aristotle's, but one that may owe something to his inspiration.

Slavery, it was argued, was 'progress': 'an integral link in the grand progressive evolution of human society', as William Henry Holcombe, writing in 1860, put it. Moreover slavery was a natural means of 'Christianisation of the dark races' (Holcombe 1986, p. 23). Slavery was assumed to be one of the means whereby the natural, debased life of the primitives could be civilised. And in this it may not be too far-fetched to see at least a touch of Aristotle's logic which defended human slavery on the basis that domestic tame animals were better off 'to be ruled by men, because it secures their safety' (Aristotle 1985, p. 68). As David Brion Davis points out:

> It is often forgotten that Aristotle's famous defence of slavery is embedded within his discussion of human 'progress' from the patriarchal village, where 'the ox is the poor man's slave', to the fully developed *polis*, where advances in the arts, sciences and law support that perfect exercise of virtue which is the goal of the city state (D.B. Davis 1986, p. 25).

If slavery then was frequently defended on the basis of 'progress', on which basis, we may ask, was it opposed? We know that reformers – like Shaftesbury, Wilberforce, Richard Bater and Thomas Clarkson – opposed the trade in slaves because they regarded it as cruel, dehumanising and the source of all kinds of social ills. One argument which they used time and again was that man had no right to absolute dominion over other men. According to Theodore Weld's influential definition, slavery usurped 'the prerogative of God'. It constituted, 'an invasion of the whole man – on his powers, rights, enjoyment, and hopes [which] annihilates his being as a MAN, to make room for the being of a THING' (Weld 1986, p. 146). In other words, humans cannot be owned as things or as property. This argument was not peculiar to Weld and Wilberforce and the other reformers in the eighteenth and nineteenth centuries. Nearly 1,400 years before Wilberforce was born, St Gregory of Nyssa made the first theological attack on the institution of slavery itself. His argument is simple: man is beyond price. 'Man belongs to God; he is the property of God'; he cannot therefore be bought or sold. St Gregory was the first to break decisively with that 'belong to and exist for' element within the Western tradition (Dennis 1986, p. 138). Yet St Gregory's argument contains a twist in its tail, for St Gregory argues that humans cannot have dominion over other humans and therefore possess them, because God gave humans dominion, not over other humans, but over the world and animals in particular. In other words, humans belong to God and are therefore beyond price but the animals, since they belong to humans, can be bought and sold like slaves (Dennis 1986, p. 137).

We are now in a position to confront the second type of slavery, namely the slavery of animals. We find, almost without exception, all the same kinds of argument used to justify human slavery also used to justify the slavery of animals. Animals, like human slaves, are thought to possess little or no reason and, like human slaves, are thought to be 'by nature' enslaveable. Animal slavery, like human slavery, is thought to be 'progressive', even of 'benefit' to the animals concerned. But two arguments are used repeatedly, and we have already discovered them: animals belong to humans and they exist to serve human interest. Indeed Brion Davis describes what is meant by a slave in a way that makes the similarity abundantly clear:

> The truly striking fact, given historical changes in polity, religion, technology, modes of production, family and kinship structures, and the very meaning of 'property', is the antiquity and almost universal acceptance of the *concept* of the slave as a human being who is legally owned, used, sold, or otherwise dis-

posed of as if he or she were a domestic animal (D.B. Davis 1986, p. 13).

You may well ask what the foregoing has to do with the issue of genetic engineering. The answer is simply this: genetic engineering represents the concretisation of the *absolute* claim that animals belong to us and exist for us. We have always used animals, of course, either for food, fashion or sport. It is not new that we are now using animals for farming, even in especially cruel ways. *What is new is that we are now employing the technological means of absolutely subjugating the nature of animals so that they become totally and completely human property.* 'New animals ought to be patentable', argues Roger Schank, Professor of Computer Science and Psychology at Yale University, 'for the same reason that new robots ought to be patentable: because they are both products of human ingenuity' (Schank 1988). When technologists speak, as they do, of creating 'super animals', what they have in mind is not super lives for animals so that they may be better fed, have more environmentally satisfactory lives, or that they may be more humanely slaughtered rather what they have in mind is how animals can be originated and exist in ways that are completely subordinate to the demands of the human stomach. In other words, animals become like human slaves, namely 'things', only even more so, in a sense, since human masters never – to my knowledge – actually consumed human slaves. Biotechnology in animal farming represents the apotheosis of human domination. In one sense it was inevitable. Failing to have respect for the proper limits in our treatment of animals always carried with it the danger that their very nature would become subject to a similar contempt. Now animals can not only be bought and sold but *patented*, that is owned, as with human artefacts, like children's toys, cuddly bears, television sets, or other throwaway consumer items, dispensed with as soon as their utility is over.

Again we are not, even at this point, as far away from Aristotle's thinking as some might suppose. For in an uncanny, prophetic part of his work, Aristotle seems to anticipate a time when human slaves would be automated, being slaves of their own nature, rather than being enslaved by nature or the will of their masters. 'For suppose,' he muses, 'that every tool we had could perform its task, either at our bidding or itself perceiving the need ... then mastercraftsmen would have no need of servants nor masters of slaves' (Aristotle 1985, p. 65). Some might argue that genetic engineering has transformed this ancient dream into a present nightmare.

Claims on Creation

I was going to call this chapter 'the discredited theology of genetic engineering' but some individuals protested that this might be read as assuming that genetic engineering *had* a theology. In fact it does, and a strong and powerful one at that. The Christian tradition, fed by powerful Aristotelian notions, has been largely responsible for its propagation. For many centuries Christians have simply read their scriptures as legitimising the Aristotelian dictum, 'existing for and belonging to'. The notion of 'dominion' in Genesis has been interpreted as licensed tyranny over the world, and over animals in particular. God, it was supposed, cared only for humans within his creation, and as for the rest, they simply existed for the human 'goodies'. According to this view, the whole world belongs to humans by divine right, and the only moral constraint on the use of animals was how we should treat them if they were the property of other humans (H. Davis 1946, p. 258).

This God – not unfairly described as a 'Macho-god', essentially masculine and despotic, who rules the world with fire and expects his human subjects to do the same – has trampled through years of Christian history, but his influence is now waning. There are many reasons for this decline, but two in particular are as follows. First, most Christians do not believe in him anymore. You will have to search high and low for any reputable theologian who defends the view that God is despotic in his power and wants his human creatures to be so as well. Second, having re-examined their scriptures, most theologians conclude that we misunderstand dominion if we think of it simply in terms of domination. What dominion now means, according to these scholars, is that humans have a divine-like responsibility to look after the world and to care for its creatures (Baker 1975). Indeed, for those of you who still hold to the 'Macho-god' version of divinity, I have some worrying news. Not only theologians (who tend in the nature of things to be either ahead or behind the times), but also church people, even church leaders, have now disposed of this old deity:

> The temptation is that we will usurp God's place as Creator and exercise a *tyrannical* dominion over creation ... At the present time, when we are beginning to appreciate the wholeness and inter-relatedness of all that is in the cosmos, preoccupation with humanity will seem distinctly parochial ... too often our theology of creation, especially here in the so-called 'developed' world, has been distorted by being too man-centred. We need to maintain the value, the preciousness of the human by affirming the preciousness of the non-human also – of all that is. For our

concept of God forbids the idea of a cheap creation, of a throw-away universe in which everything is expendable save human existence. The whole universe is a work of love. And nothing which is made in love is cheap. The value, the worth of natural things is found not in Man's view of himself but in the goodness of God who made all things good and precious in his sight ... As Barbara Ward used to say, 'We have only one earth'. Is it not worth our love? (Runcie 1988)

These are not my words. They come from a recent lecture given by the Archbishop of Canterbury, Robert Runcie. You will notice how the earlier tradition is here confronted and corrected. God is a God of love. His world is a manifestation of his costly, self-sacrificial love. We humans are to love and reverence the world entrusted to us. And lest you should think that this is just 'Anglican' theology which may at times tend to be a little fashion-able, it is worth mentioning that the latest encyclical from that rather unfashionable, undoubtedly conservative Pope, John Paul II, specifically speaks of the need to respect 'the nature of each being' within creation, and underlines the modern view that the 'dominion granted to man ... is not an absolute power, nor can one speak of a freedom to "use and misuse", or to dispose of things as one pleases' (Pope John Paul II 1988).

We have not yet brought our argument to its sharpest point however. It is this: *no human being can be justified in claiming abso-lute ownership of animals for the simple reason that God alone owns creation. Animals do not simply exist for us nor belong to us. They exist primarily for God and belong to God. The human patenting of animals is nothing less than idolatrous.* The practice of genetic engineering implicitly involves the claim that animals are ours – to do with as we wish and to change their nature as we wish. The same reason why it is wrong to use human beings as slaves is also precisely the same reason why we should now oppose the whole genetic engi-neering endeavour with animals as theologically erroneous. We have no right to misappropriate God's own.

I anticipate four objections to this conclusion which I consider briefly in turn below.

The first objection is as follows. We have always made animals our slaves. Our whole culture is based upon the use of animals. It is therefore absurd to suppose that we can change our ways.

I agree with the first part of this objection. It is true that our whole culture is based on the slavery of animals. I, for one, would like to see a root and branch cultural change. Christians as a whole may legitimately disagree about how far we can and should use animals. But one thing should be clear: we cannot own them, we should not treat them as property, and we should not pervert their nature for the sole purpose of human consumption. Genetic engineering – while part of this cultural abuse of animals – also represents its highest (or lowest) point. Because we have exploited in the past, and continue to do so, is no good reason for intensifying that enslavement and bringing the whole armoury of modern technology to bear in order to create and perpetuate a permanently enslaved species.

The second objection is that the record of Christianity has been so terrible as regards the non-human that we must surely despair of a specifically theological attempt to defend animals.

I agree with the first part of this objection. Christianity has a terrible record on animals. But not only animals – also on slaves, women, the mentally handicapped, and a sizeable number of other moral issues as well. I see no point in trying to disguise the poor record of Christianity, although I have to say that I do not quite share Voltaire's moral protest to the effect that 'every sensible man, every honourable man, must hold the Christian sect in horror' (Voltaire 1986, p. 130). All traditions, religious or secular, have their good and bad points.

To take but one issue as an example. Recall my earlier point that if we went back in history 200 years we would find intelligent, conscientious, respectable Christians defending slavery as an institution. The quite staggering fact to grapple with is that this very same Christian community came, within an historically short period of between 50 and 100 years, to change its mind. This community, which in some ways provided the major ideological impetus for the defence of slavery, was the community that helped to end it (D.B. Davis 1986). So successful indeed has this change been that I suppose among Christians today we would have difficulty in finding one slave trader, or even one individual Christian who thought that the practice was anything other than inimical to the moral demands of the Christian faith. In short, while it is true that Christian churches have been and are frequently awful on the subject of animals, it is just possible, even plausible, that given say 50 or 100 years we shall witness among this same community amazing shifts of consciousness, as we have witnessed on other moral issues, no less complex or controversial. In sum, Christian churches have been agents of slavery – I do not doubt – but they have also been, and can be now, forces for liberation.

The third objection is that genetic engineers are really good, honest, loving, generous, well-meaning people only trying to do their best for the sake of humanity, or at the very least they are no more awful than the rest of us. Why criticise what they are doing, when in point of fact *everybody* is in the difficult situation of moral compromise to a greater or lesser extent?

Again, I agree with the first part of this objection. I have no reason for doubting the sincerity, the motivation and the moral character of those who are actively engaged in genetic engineering research. One of the really sad aspects of the campaign for the abolition of the slave trade was the way in which abolitionists tended – during the time of their ascendency – to vilify their opponents by regarding them as the source of all evil. I have no desire to do the same. Indeed, what I want to suggest is that genetic engineers are really doing what they say they are doing, namely pursuing the cause of humanity according to their own lights. What I want to question, however, is whether a simple utilitarian humanist standard is sufficient to prevent great wrong. From the slave trader's perspective it was only right and good to use slaves for the sake of the masters. From the genetic engineer's perspective it is only right and good to treat animals as utilities in order to benefit the human species. I doubt whether a simple utilitarian calculation based on the interests of one's *own* class, or race or species can lead other than to the detriment of *another's* class, race or species. Once we adopt this framework of thinking, there is no right, good or value that cannot be bargained away in pursuit of a supposedly 'higher' interest.

The fourth objection is to the analogy so far drawn between human and animal slavery. Animals are only animals, it is argued. Animals are not human.

This argument, which emphasises a clear demarcation between animals and humans, whatever its merits in other spheres, is exceedingly problematic when applied to genetic engineering. After all, are not genetic engineers involved in the injection of *human* genes into non-human animals? According to a recent report, Dr Vernon G. Pursel, a research scientist at the US Department of Agriculture (USDA) research facility in Beltsville, responded to a recent move by various humane agencies and churches against genetic engineering by saying, 'I don't know what they mean when they talk about species integrity.' He went on to make the following most revealing statement, 'Much of the genetic material is the same, from worms to humans' (Pursel 1986, p. 5).

I find the above statement by Dr Pursel revealing, precisely because it supposes what transgenic procedures must implicitly accept, namely that there is not a watertight distinction between humans and animals. Some may think that this is an argument in favour of treating animals in a more humane fashion, and so in a way it should be, but the argument is used to the practical detriment of non-human creatures. Here we have a curious confirmation of the anxiety that besets bystanders like myself, for the question that must be asked is this: if the genetic material is much the same – from worms to humans – what is there logically to prevent us experimenting upon humans while accepting its legitimacy in the case of animals? Indeed genetic experiments on humans are not new, and neither is the view that there should be a eugenics programme for human beings. This view has received strong support from Christians at various times. One respected Christian writer in 1918 made clear that:

> The man who is thoroughly fit to have children, and who either through love of comfort, or some indulgence of sentiment, refrains from marriage, defrauds not only himself and his nation, but human society and the Ruler of it ... But the man or woman who knowing themselves unfit to have healthy children yet marry, are clearly guilty of an even more serious offence (Gardner 1918, pp. 188–9).

This writer does not just advocate these moral imperatives as personal guidelines, rather he seeks to have them enshrined in law:

> The only kinds of legislation *for which the times are ripe* seem to be two. In the first place, marriage might be forbidden in the case of those mentally deficient, or suffering from certain hereditary diseases. And in the second place, much more might be done at present in the way of providing cottages in the country, and well-arranged dwellings in the town, and by encouraging in every way the production of healthy children (Gardner 1918, pp. 188–9).

This work was entitled *Evolution in Christian Ethics*, published in the prestigious Crown Theological Library series, and written by Percy Gardner. Gardner's view was straightforward: only those who are fit have the right to propagate the race. The wellbeing of the race was, as he saw it, threatened by the First World War because only the 'weaker, and especially those whose vital organs are least sound, we retain at home to carry on the race' (Gardner 1918, p. 190).

Gardner had to wait another 15 years before his ideas reached their fullest and most persuasive expression in the work of another writer, a political philosopher, of immense influence:

[The state] must see to it that only the healthy beget children; but there is only one disgrace: despite one's own sicknesses and deficiencies, to bring children into the world, and one's highest honour to renounce doing so ... [The state] must put the most modern medical means in the service of this knowledge. It must declare unfit for propagation all who are in any way visibly sick or who have inherited a disease (Hitler 1974, pp. 367–9).

And according to this view:

[The state's philosophy] of life must succeed in bringing about that nobler age in which men no longer are concerned with breeding dogs, horses, and cats, but in elevating man himself, an age in which the one knowingly and silently renounces, the other joyfully sacrifices and gives (Hitler 1974, pp. 367–9).

These views are taken from the well-known work, *Mein Kampf*, and the author is, of course, Adolf Hitler.

Some may object that the analogy here breaks down. After all, Hitler would hardly have approved of infecting Aryan blood with the genes of animals or, more accurately, allowing Aryan genes to be 'wasted' on animals. He was hardly in favour of 'hybrid humans' – as he called the children of mixed marriages – so he might well have had a certain disdain for the very idea of transgenic animals. And yet we cannot dismiss the fact that Hitler popularised, indeed did much to develop, a medical science which aimed at 'preserving the best humanity' as he saw it. What is more, his ideas of genetic control exercised through force, coercion and legislation are by no means defunct. Indeed the notion of creating a 'super animal' is faintly reminiscent of the Hitler doctrine of creating a 'superior race'.

Some may still feel that human genetics, and genetic engineering with animals are two quite separate things. Some may think that I am being simply alarmist. However, *Mein Kampf* is, in my view, a much more important work of political philosophy than its detractors allow – but that is beside the point. The point is that I can find no good arguments for allowing genetic experiments on animals which do not also justify such experiments (or genetic programmes) on human beings. I am alarmed by the way in which we have simply failed to recognise that animal experiments are often a precursor to experiments on human beings. Even in current established practice, animal experiments frequently precede clinical trials on human subjects. We should not be oblivious to the fact that the century which has sustained the most ruthless scientific research on animals is also the century that has seen experiments on human subjects as diverse as Jews, blacks, prisoners of war and human embryos. If 'much of all genetic material is the same – from worms to humans', as Dr Pursel maintains (Pursel 1986), what real difference does it make whether the subjects of such experiments are animals or humans?

There is one important sense in which Dr Pursel was right: in addition to the nature appropriate to each individual species, there is a nature which is common to all human and non-human animals. This realisation alone should make us think twice about genetic engineering. Animals, it is sometimes supposed, are simply 'out there', external to ourselves like nature itself. Likewise, it is thought, what we do to animals does not really affect *us*. In fact, humans are not just tied to nature, they *are part of nature*, indeed inseparable from nature. Because of this there is a profound sense in which we cannot abuse nature without abusing ourselves (Mar Gregorios 1987). The genetic manipulation of animal nature is not just some small welfare problem of how we should treat some kinds of animal species; it is part of a much more disturbing theological question about 'who do we think we are' in creation, and whether we can acknowledge moral limits to our awesome power, not only over animals, but also over our own species.

At the beginning of this chapter, I invited you to imagine the Old Major in Orwell's *Animal Farm* addressing his fellow animal comrades, complaining that their state was none other than 'misery and slavery'. You may recall that, a little provocatively, the Old Major thought that the answer was the abolition of 'Man'. In one sense the Old Major was right. We need to abolish what St Paul calls the 'old man' which is the real slavery, namely the slavery to sin (see, for example, Romans 6:6). St Paul understands Christian discipleship as 'dying and rising with Christ' in baptism whereby our old selfish natures are transformed. Demythologised a little, what St Paul might have said is that we must stop looking on God's beautiful world as though it was given to us so that we can devour, consume, and manipulate it without limit. I look forward to the final death of the 'old man' – of which St Paul speaks – both in myself as well as in other human beings. Then, and only then, when we have surrendered our idolatrous power – nothing short of tyranny – over God's good creation, shall we be worthy to have that moral dominion over all which God has promised us.

Indeed, the Old Major spoke to the assembled animal throng of a dream he had had about animal liberation. It took the form of a song, with 'a stirring tune, something between "Clementine" and "La Cucuracha"' (Orwell 1961, p. 14). Here are the first two stanzas:

> Beasts of England, beasts of Ireland,
> Beasts of every land and clime,
> Hearken to my joyful tidings
> Of a golden future time.
> Soon or late the day is coming,
> Tyrant man shall be o'erthrown,
> And the fruitful fields of England
> Shall be trod by beasts alone.
> (Orwell 1961, p. 14)

Acknowledgements
I am grateful to Joyce D'Silva for her kind research assistance during the preparation of this chapter. Also thanks are due to Dr Paul A.B. Clarke and Dr Peter J. Wexler of the University of Essex who kindly read an earlier draft of this chapter and gave me the benefit of their comments.

18 Pigs *Will* Fly

RICHARD RYDER

Speciesism

In this chapter I wish to look at genetic engineering as it applies to non-humans, and at humankind's new-found power to mix the species to produce new ones. Genetic engineering highlights the difficulty of defining what a species is. What an extraordinary state of affairs: one species is now in a position to create brand new species! Not satisfied with natural evolution, humans believe we can improve upon it, mixing up genes in exciting new combinations to produce animals that are bigger and better than 'God's own handiwork'. Only a few hundred years ago people would have been burned at the stake if they had suggested such a blasphemy!

For centuries there has been a horror of transgressing the natural boundaries between human beings and other species. Humankind has had to assert its difference from the other animals. Sexual intercourse, for example, between humans and beasts was regarded as the blackest sin. Why did people so much fear the blurring of the species boundaries? The answer is that most of us are 'speciesist' snobs: we like to feel that the human animal is in a class apart from the others – that humans are of a superior status to the non-human creation. Furthermore, our definition of the genetic separation of species has satisfied the human need to find order and to classify our environment into convenient pigeon-holes. This has given us a sense of security.

From Aristotle onwards many philosophers have taken comfort from seeing nature as a fixed hierarchy of genetically separate species each with their own natural calling. Gradually, from Copernicus and Galileo to Charles Darwin, the concept of the uniqueness and centrality of humankind has been eroded. Yet we still adhere to the primitive morality of speciesism which insists upon an artificial line being drawn between our own species and the others. 'Do unto them what you would *not* do unto us', has been the guiding rule.

One benefit of genetic engineering is that it will undoubtedly blow a hole in this conventional hypocrisy. The watertight, or 'genetight', boundary between species (widely and erroneously regarded as far more absolute than really it has ever been) has now been perforated by the injection of genes from one species into the embryos of others to produce new creatures.

I am glad to say that, at the psychological level, genetic engineering is likely to give speciesism a very bad name in the future. It spotlights the absurdity of our species-based morality. Human growth hormone genes, yes *human* genes, have already been injected into the embryos of pigs. The aim was to produce bigger and juicier pork chops. But wait a minute. This would mean *eating* human genetic material! It might be only a minute proportion of the chop, but all the same, would it not be partial cannibalism?

This speculation raises interesting questions. How many human genes make a sufficiently human creature to have human rights in the eyes of the law? How many human genes can you give a humanised pig before you feel obliged to send it to school rather than to the slaughter house? Moreover it is now theoretically possible to start with human embryos and inject little bits of monkey or dog or plant or insect into them. So, when a slightly 'monkeyfied' or 'dogified' baby is born, will it eventually be educated, or put in a cage and vivisected?

It is no use for scientists to argue that this has not happened. The point is that the technology is now there to do it, and the history of science strongly suggests that if things *can* be done then sooner or later they *will* be done. Such is the Frankenstein Revolution we are facing. Are we planning a future with 4 metre long pigs, and cows the size of elephants?

I think that a traditional Christian point of view raises many objections to genetic engineering, not least the objection to man playing at being God. A Vatican spokesman has described the prospect of the interbreeding of humans with non-humans as 'a satanic attempt to destroy every essence of God in the Universe, destroying His likeness, which is Man' (IPPL Newsletter 1987). From my own more hedonistic moral viewpoint, genetic engineering is, in itself, neither right nor wrong. Like many increments in knowledge it is a discovery which can be used either for good or for ill. In itself it is neutral.

Sentientism

Let me briefly explain my own moral position which is one that could, for lack of a better term, be labelled 'sentientism'. That is to say, I hold it wrong to cause pain or distress to any sentient being, unless it is with their agreement, or unless it will bring unquestionable benefit to that *same* individual if he or she is unable to give informed consent. Sentientism is different from utilitarianism, of which I have sometimes been accused, because it rules out any question that the suffering of one sentient can be justified in terms of benefits to others.

In my opinion ends cannot justify means if the means themselves cause involuntary suffering. This is approximately the animal rights position as expounded by Tom Regan (1983) and others. However, like Peter Singer (1977), I place my emphasis upon the criteria of pain and pleasure rather than upon vaguer concepts such as the 'inherent value' of human and other animals. I differ from Singer in stressing the importance of the individual sentient, whether human, non-human or partially human. For me it is never possible to justify the suffering of one for the pleasure of another.

It is the suffering which might be caused to sentients, whether wholly or partially human, or entirely non-human, which must concern us. There seem to be three main aspects of genetic engineering which could cause suffering. First, suffering could be caused to sentients in the procedures leading up to the production of actual transgenic animals. Secondly, the transgenic animals themselves may suffer, either through hereditary factors (such as the severe arthritis found in the Beltsville 'humanised' pigs) or through their subsequent treatment. Thirdly, suffering may be caused through the products of genetic engineering upon the environment, such as those lucidly hypothesised by Michael Fox [see Chapter 5] and by Andrew Lees [see Chapter 13].

We have known for years of the congenital miseries of so-called 'purebred' dogs, horses, cats and farm animals. Humankind's selective breeding *within* species has already produced suffering enough in attempts to breed faster racehorses, leaner meats, more absurd looking pets and quicker profits.

Nearly 200 diseases of genetic origin, such as hip dysplasia, are already identified in highly inbred dogs. Genetic engineering, which vastly accelerates the way in which humankind can alter the natural shape of animals, is almost certain to worsen this lamentable situation. With investment in genetic engineering running to mountainous sums, scientists are going to be under inestimable pressure to produce results – regardless of the suffering involved.

I am not saying that genetic engineering will never bring benefits. Cultures of genetically engineered bacteria for example can mass produce human insulin for diabetics. Human genes involved in the production of Factor IX, which some haemophiliacs need so desperately, have been inserted into sheep to produce adult sheep who secrete Factor IX in their milk [see Chapter 3]. Genetically manipulated microbes can be produced which protect plants from disease. I am arguing that where genetic engineering produces suffering, in humans or in non-humans, then it is wrong. As sentientists we must concentrate simply upon this objective: to reduce pain and distress in all sentient individuals.

RSPCA Recommendations

Why have we in the UK been so slow to sound the alarm on genetic engineering? The first new transgenic animals were made in 1982 when Drs Palmiter and Brinster produced 'rattish' mice in the USA (Palmiter *et al.* 1982; Palmiter and Brinster 1986) [see Chapter 3]. Why has it taken six years for the implications of the Frankenstein Revolution to reach us in the UK animal welfare movement? I think the answer is that the UK public does not yet fully understand what is going on. Perhaps the UK media is also a little out of touch. The UK public may not be aware that many *billions* of dollars have been poured into this field in the search for new and highly profitable products. They may not know that in the USA it is now possible to patent genetically engineered animals. It has been our ignorance which has caused our silence.

The commercial and political implications of genetic engineering are colossal. We must at once find ways to safeguard the interests of sentients at all three stages of the biogenetic industry – research, production and release. This is why the RSPCA has made the following recommendations.

Under the Animals (Scientific Procedures) Act (1986) any project licence granted to a laboratory for altering the genetic make-up of any vertebrate must not be allowed to exceed the 'mild severity' category as defined for project licence applications. This would ensure that should there be any suffering involved, the animal would be 'euthanised' at an early stage.

The British Home Office should make public the details of genetic engineering research, such as the number of animals born with defects or not surviving to birth, the percentage of survivals and the severity banding of each project, as part of the *Annual Statistics of Experiments on Living Animals.*

Genetically engineered animals should only be allowed out of Home Office licensed premises if the genetic manipulation is not in any way detrimental to the health or wellbeing of that animal under any circumstances.

The Health and Safety Executive (HSE), in considering the planned release of transgenic animals, should liaise closely with the Home Office to ensure the continued welfare of the animals concerned.

In formulating guidelines for the genetic manipulation of sentients, consideration must be given to the possible threat to the animals subjected to such experiments and to the environment in which they live.

The HSE Subcommittee which considers transgenic manipulation should include representatives from the Farm Animal Welfare Council (FAWC), the Animal Procedures Committee (APC) and an external animal welfare representative [see also Chapter 1].

The Local Genetic Manipulation Safety Committees, set up under the HSE [see Chapter 23] at establishments conducting research on transgenic animals should include an external animal welfare representative.

The proposed European Directive on Planned Release [see Chapter 23] should be cross-referenced to European Community (EC) directives concerning animal welfare.

Finally, the RSPCA is concerned that the enormous financial incentives involved in the patenting of transgenic animals will lead to the over-riding of welfare considerations. Therefore the RSPCA is opposed to the patenting of animals.

Conclusions

My own attitude towards genetic engineering is one of ambivalence. For nearly two decades I have been predicting the mixing of human and non-human genetic material and warning that it will create a crisis in our moral outlook.

When we create new species containing human genetic material then what is the moral and legal status of such creatures? The Beltsville pigs bridge the artificial conceptual gulf between human and non-human. This may merely be a slender crossing but wider bridges no doubt will follow which will make a nonsense of our species boundaries and species-based morality. Either we will treat these 'humanised' creatures in the same tyrannical way in which we have treated non-humans for thousands of years (in which case there will be less reason than before why we should not also exploit human beings in the same way) or we must take up our moral duties to all sentient beings and stop wilfully inflicting suffering upon any of them.

Uncontrolled gene splitting could surely be as dangerous as uncontrolled atom splitting. Genetic catastrophes, if no limits were drawn, could be as likely as nuclear catastrophes. In addition I fear that genetic engineering might foster an increasingly exploitive and materialistic attitude towards our environment. Surely this revolution in science is likely to make humankind even more conceited than it already is. Conceit and pride often come before a fall.

It is no good for scientists to be complacent – they can foresee the ultimate consequences of their own work all too rarely. Rutherford, shortly after splitting the atom, said he could foresee no possible practical applications for this new technology!

I believe we are on a slippery slope where, as Edward Lee Rogers reminds us [in Chapter 21], new technologies in the absence of proper regulatory legislation become imperative: what *can* be done *must* be done and *will* be done.

Increasingly, public mistrust of science is, thank goodness, being openly shared by some scientists. But the promises of the multinational companies, and the career prospects that they offer to scientists in the field of genetic engineering, are understandably persuasive. Science and commerce cannot be left to regulate themselves: the competition and the incentives are too strong. Genetic engineering, like all things with great potential for good or for evil, must be a matter for open public debate and independent statutory control. There is no room for secrecy. Proper international legislation to protect non-humans and humans alike from pain and distress is required now, before the genetic engineering bandwagon rolls any faster.

19 Blueprint for a Humane Agriculture

PETER ROBERTS

Eminent scientists have developed complex methods to analyse life itself. We are told that life can be understood in terms of the sequence of bases in the DNA molecule – a double-stranded helix held together by many millions of links which occur in ordered sequence. It is suggested that the secrets of nature have been reduced to their essentials, unlocked and thus explained.

I neither want, nor am I able, to elaborate upon the complex methods used by these scientists. In this chapter I wish to take an overview – a view of synthesis rather than one of analysis. I shall consider the development of our dependence on farm animals, how we have influenced their evolution, and importantly, how the new science of genetic engineering will affect their future evolution.

In terms of the power which the experimenter has over other life-forms, we can say that selective breeding is to genetic engineering what the abacus is to the computer. We should remember also what use we have made of the power of selective breeding – the deformed face of the snuffling pekinese and the giant baby chicken of the broiler industry: creatures developed for our frivolous sense of fashion, or for short-term advantage in the market place. And now, with such a history, we throw away the abacus and take up the computer! What lies ahead?

The Place of Ethics

I will consider first the effect of the new knowledge of genetics on our attitudes and ethics.

The reductionist view of life on earth ignores the function of mind, which I believe lies behind all the structures of nature. Nothing that has been discovered denies that view. All too often the discovery of the structure of DNA is taken as an implied proof of atheism. It is such a fashionable view that one can almost hear some of the experimenters thanking God that they are atheists!

The reductionist view is often thought to exclude mind – an extraordinary exclusion since the sequential arrangement within the DNA molecule does itself suggest the intelligent direction of mind.

Look at it this way: a child discovers that a book is not merely a block of paper, but that it is made up of pages on which are imprinted an almost infinite permutation of 26 letters, some of which always have a preference for combining with others into certain groups called syllables and words. This discovery no more denies the reality of the author of the book than the reductionist's explanation of life denies the pre-existence of mind. If the mind persists beyond a physical terminus perhaps it also preceded it – as a totality – the author of the ordered creation of nature into which it then immersed itself. My excuse for this metaphysical diversion is to expose as false the view so often held that the discovery of the basis of heredity renders religion obsolete, and with it the need to act with due regard to ethics in our behaviour towards the rest of the creation.

Balanced Farming

I will turn now to a consideration of the science and practice of food production rather than metaphysics or philosophy.

For perhaps forty thousand years, the human race has exploited animals for its own purposes, first as a hunter, then as a shepherd and more lately as a tyrant. Early in this century an ecologically sound system of rotational farming was evolved that had, built into it, pest control, weed control and disease control. It was a system in which all wastes were returned to the soil for recycling, in which there was no chemical pollution nor organic pollution of natural resources. It was a system in which the animals and poultry were allowed to build up a natural relationship with each other in the herd and flock and in which they were given a tolerably square deal before they were killed. During this time selective breeding was carried out, usually within breeds, and first-crosses were developed to exhibit what was called 'hybrid vigour'.

This sound system, in which livestock were oriented to the natural world, was characterised by certain taboos built into farm tenancy agreements: no land should have the same crop more than two years running, except pasture; no hay or straw should be sold off the farm unless it was essential – in which case the equivalent fertility had to be replaced by a succeeding crop of green manure no straw must ever be burnt, and, unwritten in any tenancy agreements but nonetheless observed, no pregnant animal should be sent for slaughter.

Then came the genetic selection of livestock as a science. More complex hybrids were developed that could out-perform anything previously achieved by purebreds or first-crosses. At the same time the science of animal nutrition and antibiotics took a leap forward, which removed certain limiting factors.

These limiting factors had prevented farmers from keeping too large a population of livestock on their premises. Nature had used disease to prohibit too large a flock of egg-laying chickens, for example. Now, with the help of antibiotics, farmers could cheerfully increase their flocks from 1,000 birds to 6,000, and later to 60,000.

'Factory Farming'

The chemist, the nutritionist and the geneticist enabled the greedier farmers to grab the markets previously supplied by all their colleagues. Under this new regime, animals could no longer be allowed to live under natural conditions and so were put inside huge, windowless sheds, often in cages.

Slowly, all too slowly, the public, speaking with a corporate voice through the farm animal welfare movement, let its displeasure be known, and put pressure on the politicians to limit the excesses of the 'factory farmers'. But, as always with politicians, there was compromise upon compromise, and there was also the need for the courts to deliberate endlessly on the meaning of some intentionally vague phrase such as 'unnecessary suffering'.

On behalf of Compassion in World Farming (CIWF), I took a court action against the monks of Storrington Priory who showed their concern for the descendants of the biblical fatted calf by imprisoning some 650 white veal calves in individual pens ('crates'), each less than 2 feet wide, and in which they were unable even to turn round and quite unable to groom themselves properly. If we had won that court case we would have proved that use of the white veal crate constituted cruelty in English law. As it was we lost the case, but we were able to demonstrate that the voluntary codes of welfare practice were widely ignored and were no substitute for mandatory regulation. Fortunately we now have mandatory regulation in Britain, and I understand that by the end of 1989 the European Community (EC) is likely to emulate Britain in this respect by publishing minimum standards for calves.

The current position with other species of livestock can easily be summarised. With regard to chickens our legislators have ignored the advice of their own Committee of Enquiry and their own Agriculture Committee, and adopted the commonplace battery cage size as a legal minimum, bowing to our European masters who have decided on a cage space per hen of 450 cm^2 – a lot smaller than an average shoe box. In reality 450 cm^2 is the physical space occupied by a hen at rest. If the hen should want to preen herself or even ruffle her feathers, she must invade the space of her cage companions. Of course, if she were a wild bird she would, by law, need space within her cage to spread her 32 inch (80 cm) wing-span; by present standards, 32 inches is the cage length of, not one, but eight chickens.

The majority of our breeding pigs spend the whole of every pregnancy lying in a dry sow stall or tie-stall without the opportunity for exercise. They adapt to their close confinement in a manner which resembles in humans the development of chronic psychiatric disorder.

I have described 'factory farming' up to the present time and, as the author Alan Paton, wrote of South Africa in his autobiography: 'For today one despairs; for tomorrow one hopes' (Paton 1988). The new science of genetic engineering has the potential both to realise, or to shatter, the hopes for improvements in the welfare of farm animals.

Public Concerns

The public has not been slow to express its concern about the power of genetic engineering. The researchers are promising us rich dividends if they are given a free hand, but the public recalls hideous and offensive scientific research by other scientists in recent times. To give just three examples of such research: a farm animal's udder was amputated and then surgically attached to the same animal's neck with a direct blood supply from the carotid (neck) arteries; a dog was given an additional head surgically removed from another dog (Ryder 1983); a patent was requested for a device in which to maintain life in the severed head of a monkey – or human – which, and I quote, 'may still use its vocal cords' – presumably to scream (US PTO 1987; see also NAVS 1988).

In a British television programme in the *Nature* series, Dr Chris Polge of Animal Biotechnology Cambridge Limited said, 'The main qualm I have is man and what use he makes of the technology. I have the same qualms about nuclear energy as I have about genetic engineering' (*Nature* 1988). There is some analogy between nuclear energy and genetic engineering.

The dangers of the build-up of nuclear weapons were so great that they were addressed not on a national level but on a global level. We have the Non-Proliferation Treaty for example. The dangers associated with genetic engineering are insidious in the sense that covert experimentation may only be admitted to if the result is successful. If, by some unfortunate circumstances, a new pathogenic organism emerges and escapes to wreak havoc in the environment, it may not be traceable to its inventor. The question has already been raised whether certain recent, disastrous diseases such as Dutch Elm Disease, distemper in seals and AIDS in humans are the result of our interference with nature (see, for example, NAVS 1987); this may or may not be true. Genetic engineering has the depressing quality of creating both utopian and dystopian extremists. But who is to say where the line is to be drawn between dystopian extremism and a proper concern? In the case of the regulation of genetic engineering, the usual political habit of compromise may mean the elimination of half a dozen species, perhaps our own included!

A Committee for Safety and Ethics

In the early 1970s there were many expressions of concern from the public, and from scientists themselves, about the safety of laboratory work using micro-organisms. Working parties were set up both in the UK and in the USA and reports were published. In the UK the Department of Health and Social Security (DHSS) and the Department of Education and Science (DES) formed advisory groups, namely the Dangerous Pathogens Advisory Group (DPAG) and the Genetic Manipulation Advisory Group (GMAG). In the 1980s, both the DPAG and the GMAG were brought under the auspices of the Health and Safety Executive (HSE) of the Department of Employment (DOE) as the Advisory Committee on Dangerous Pathogens (ACDP) and the Advisory Committee on Genetic Manipulation (ACGM). The DPAG, GMAG, ACDP and ACGM have been dominated by scientists involved in genetic engineering, on the principle that self-regulation is the best approach. The same is true of the National Institutes of Health (NIH) Recombinant DNA Advisory Committee (RAC) in the USA.

But can an advisory group be a controlling authority? At least in the USA, under the Freedom of Information Act, the proceedings of the RAC are available for anyone to see. In the UK, the members of the genetic manipulation and dangerous pathogens advisory committees operate under the Official Secrets Act.

There is one other extraordinary thing about these government advisory committees – they are exclusively concerned with safety. No one is denying that safety is tremendously important, but what about ethics? It is not within the terms of reference of these advisory committees to question the ethics of the experiments they assess. Should we not be questioning the ethics of attempting to produce a wingless chicken or any animal with a new genetic construction?

We constantly assume that if a line of research holds any hope of being in the immediate interests of mankind, no matter how frivolous or short-term, then it is acceptable, even if that interest is confined to a passing fashion or short-lived financial advantage in the market. We only question the research in terms of the interests of the species we are tinkering with – for example, will it be more susceptible to disease – when commercial interests may be adversely affected. But what about the welfare of the species involved?

To give customers an advantage in the market, an experimenter is quite ready to implant some new growth potential in an animal, cheerfully accepting the genetically based imperfections that it may induce in the animal, which may have to be controlled by pharmaceutical drugs.

Who will control this new science of genetic engineering and determine what is acceptable and what is not acceptable? Is there any indication that the experimenters themselves can control the excesses of genetic engineering? The history of agricultural research, animal breeding and biomedical science during the last half-century does not inspire much faith in the sanity or sensibilities of many of the experimenters.

Can control be exercised through the parliamentary political process? The history of political regulation and control over the same period does not foreshadow any contructive action other than on a 'too little and too late' basis.

Can the church be relied upon to influence control over the use of genetic engineering? The Archbishop of York has recently called for strict international rules to govern the use of genetically engineered organisms. However, apart from this initiative and those of other notable individuals, such as the Reverend Dr Andrew Linzey [see Chapter 17], the church has shown a marked reluctance to get involved in the ethics of science.

So, to whom shall we turn for some guidance, control and protection? I believe that there is a need for an international authority, like the United Nations, to consider both the safety and the ethical aspects of any proposed line of genetic manipulation. The experimenters should not be on such a panel; their function, and the function of other interested parties, should be to make submissions and to provide evidence, not to pass judgement.

The proposed international authority should have powers to require genetic engineering research organisations to deposit financial bonds that would be forfeited in the event that their research or products caused damage to health or damage to the environment. These bonds would need to be very large, perhaps measured in multimillions. The size of these bonds is an indication of the potential dangers these experiments pose to the environment and public health.

The Cornucopia Ignored

The benefits of genetic engineering are immense, we are told. For example it is claimed that the injection of genetically engineered bovine somatotropin (BST) into dairy cows will increase their milk yield by 20 per cent, and the use of genetic engineering to create transgenic livestock will result in a 30 per cent increase in their growth.

What if some procedure could offer a 500 per cent increase, with no danger of side-effects to ourselves or to the environment? Such technology already exists. It involves going straight to the plant crop for our food instead of putting it through the digestive tract of an animal.

It is now 20 years since my wife and I launched the first meat-alternative product onto the British retail market, the first textured alternative to meat, derived from the soya bean. Even in those early days, and the product has come a long way since then, it tasted like meat, looked like meat, it had the nutritional value of meat without any of the toughness, and it contained no gristle. We sold it openly, labelled it truthfully, and it found a ready market. Although our soya bean product met with its share of ridicule, for every 'clever Dick' ridiculing soya protein there were a hundred relishing it in pies, burgers and stews, often without even knowing they were eating it. The wife of a famous single-handed round-the-world yachtsman told me that the first time her husband ever complimented her on her shepherd's pie was the day she made it with this textured soya protein.

Meat Factories

Those attending a cattle breeders conference in Cambridge, England, in 1988 were somewhat stunned when they were told that within 15–20 years the production of meat need no longer involve the rearing and slaughter of animals (Ridley 1988). In this case they were not being told about textured soya proteins, but about genetically engineered meat, not grown in the stockyard but cultivated *in vitro* (in glass dishes) in commercial laboratories, from cells selected from rump steak or chicken breasts.

At one stroke the 'factory farms', the stockyards, the auction rings and the slaughter-houses would be relegated to the history books along with other of our savage customs.

This may very well happen, yet I cannot see why it should when the technology already learned with soya beans can be extended readily, and applied to other crops more suitable to the climate of the particular country. In the UK, for example, extrusion technology can be applied to the field bean, the pea crop, the white lupin and the sunflower. Both the use of crops as meat substitutes and the culture of genetically engineered meat *in vitro* could lead to the emancipation of livestock.

The application of genetic engineering for the further exploitation of animals is not justified when we already have the power to release them from their present role altogether.

Soil Fertility

One of the objections raised against the cultivation of crops such as soya beans and field beans as meat substitutes is that farmers would become more dependent on chemical fertilisers. Yet all that the soil requires for its fertility is to receive the wastes of the crops grown, whether they have gone through animals or humans or have merely been composted.

We shall eventually tire of taking our holidays bathing in our own sewage at the seaside, whether off the Isle of Wight or Sicily. The only logical way to dispose of such wastes is to develop mechanical handling devices to compost it together with municipal waste, after the extraction of bottles and cans. If particular areas have special fertiliser requirements, we can turn to the sea for an endless supply of minerals, including an extensive ancillary range of trace elements.

I believe that the use of crop growth stimulants based on highly nitrogenous materials, such as ammonium nitrate, will be drastically reduced because they pollute underground waters. Perhaps their use will be rendered obsolete altogether as the geneticist implants nitrogen-fixing genes from legumes into cereals and root crops.

Land Use

Whether the new 'meats' come from textured crops or from meat laboratories, one thing is certain: we shall in future be able to feed the world's population with a wholesome, nutritious and varied diet using only a fraction of the present cultivatable land.

Will animals disappear from the countryside? Well, undoubtedly, there will be some eccentrics who will still insist on killing an animal to eat its body – the cranks of the 21st century, but we shall start to look upon our present breeds of livestock not as sources of food but as tools for ecological management. For instance the most economic tool to keep broad-leaved forests free of undergrowth is a herd of pigs. Sheep or goats will be needed to maintain areas where it is desirable to retain hill grassland rather than woodland or scrub. I hope we shall learn to delight in animals for their own sakes, rather than for our own. We shall need to pay wardens to look after them and their environment even if the use of such land changes from food production to leisure.

Why alienate ourselves further from nature when we know that our happiness depends on harmony with nature? The choice is ours. Let us hope we make the right choice.

Discussion IV

Anon: Richard Ryder, you gave two examples of experiments which involved the insertion of a single human gene into farm animals. One was the 'Beltsville pig', into which a human growth hormone gene was inserted in the hope of producing a larger pig. The other was the insertion into sheep of a gene which codes for Factor IX, in order to obtain a blood-clotting agent for the treatment of haemophilia in humans. In both cases the species barrier has been crossed and a human gene has been put into the animal. In your view, is one of these experiments wrong, or both, or neither?

Richard Ryder, RSPCA: I hope I made it clear that in my view neither experiment is in itself wrong. An experiment is only wrong if it causes suffering, either to the animals involved in the research, or to others.

Anon: It disturbs me very much that some contributors defend the use of genetic engineering on animals providing it does not make them suffer. But suffering also includes mental suffering and psychological suffering in the sense that if we, for example, implant genes of one species into an animal of another species, it is bound to change the character and personality of the animal. Why do scientists discount the psychological and mental suffering of transgenic animals? Does it mean that our scientists do not believe that animals also have emotions, individual personalities and a sense of existence? Are we going to make the same mistake as our ancestors used to make until very recently, that animals have no souls?

Edward Lee Rogers, Foundation on Economic Trends: Richard Ryder, you have stated that in your view the moral question concerns suffering, and yet you also mentioned that you have doubts about the propriety of crossing the barrier between species. I think we have to start by questioning what we mean by species. I have had many conversations with biologists who think that our definitions of species are somewhat arbitrary.

Richard Ryder, RSPCA: I consider that genetic engineering not only undermines conventional species-based morality, but it actually undermines the definition of species itself. One of the conventional definitions of species is that if two different groups of animals cannot interbreed they are of different species. A slightly different definition is that if they can interbreed but cannot produce viable offspring they are of different species. We know that in the case of lions and tigers this definition does not hold, because if they are brought together (they are normally geographically separated by an ocean), they can interbreed and produce viable offspring which themselves can breed. I think that genetic engineering is going to highlight the fact that the whole definition of species is a nonsense and make us question our attitudes towards other sentients that happen to look different to ourselves.

Edward Lee Rogers, Foundation on Economic Trends: The Foundation on Economic Trends challenged the validity of the United States Department of Agriculture's (USDA's) transgenic animal work (for example, the creation of the 'Beltsville pig') in the courts under the National Environmental Policy Act (NEPA). We were shocked to discover that the courts did not consider that the suffering of animals was encompassed under the remit of the 'human environment', a term used in the NEPA. In my opinion, if we care about animals, including compassion for the suffering of experimental animals, then they are part of the human environment.

Mark Cantley, CUBE: I want to make a critical observation concerning Richard Ryder's presentation. While I totally share the emphasis on breaking down the speciesist attitudes of human beings, there was an innate and unmentioned cultural bias in the presentation in that all of the materials cited were very much from Western European civilisation and culture. In Confucianism, Hinduism and many of the philosophies and religions of the Far East, you would not find this degree of speciesist separation – of man's dominion over the beasts of the field. Having made that observation, this absence of speciesism does not seem to render the Eastern cultures immune to racism and its consequences. I think we should recognise the insularity of the assumptions upon which much of Richard Ryder's presentation was based. This is a good argument for internationalising the public debate about genetic engineering, to which you refer.

Richard Ryder, RSPCA: I accept that it is mainly the Judaeo-Christian tradition which has encouraged over-speciesism. Of course, whilst other major religions preach less speciesism, in practice people living in countries where these religions are dominant are also very speciesist. However, it is in countries with a Judaeo-Christian tradition where most of the transgenic manipulation research is taking place.

Bryony Cobby, RSPCA: Peter Roberts suggests that we could fertilise the land with unlimited resources from the sea. I am concerned that this is precisely the attitude we used to have towards the 'unlimited' resources of tropical hardwoods 100 years ago.

He also suggested that if we were to stop using sheep and goats for food, they could be used to maintain pastures. If we do not need sheep and goats for food, why should we maintain the pastures at all? Should we not encourage the regrowth of our natural woodlands?

Peter Roberts, CIWF: The potential for fertilising the land with resources from the sea is an area about which little is known. A considerable volume of organic manure can be extracted from seaweed, for example. I think that this would be a far better future source of fertiliser than synthetic chemicals.

Although I think there is a place for the development of natural forest, such as broadleaf forest and mixed forest, I believe that there are considerable tracts of upland grazing which the British public would not like to see reverting to forest.

We need sheep and goats as ecological tools because, without them, pasture, as we know it, would disappear. I do not see why this is in conflict with their right to lead a reasonable life. It does not suggest their exploitation – certainly not on the level to which they are exploited for food!

I would like to make a further suggestion about what we should do with farm animals when we stop eating them. We could use selective breeding to adapt them back to a more feral, original type, so that they could survive with the minimum human interference. I have a vision of the leisure parks of the future, where, when children point to wild sheep and pigs, their parents will say: 'Yes, we used to eat them once.'

Professor John Webster: To create something which makes us feel good without considering the implications for other species is carrying out moral philosophy in a vacuum. As a veterinary surgeon, one of my major concerns at the moment is the welfare of New Forest ponies and Dartmoor ponies. These ponies have very little value other than as meat exports because we are not prepared to eat them ourselves. Therefore, we have greater welfare problems with most of these animals, which are there simply as tools for the land, than those animals which have a significant cash value.

Peter Roberts, CIWF: Most of the New Forest ponies are destined for meat and are not kept as ecological tools. I am not suggesting an increase in the exploitation or the neglect of livestock. I am suggesting that, after we have stopped exploiting animals for food, we have a responsibility to look after them. For example, whether or not there was any value in their wool, sheep would still need to be sheared annually for their own welfare.

Anon: It is a general misconception that if we leave animals alone and do not interfere with them they will become our competitors. Have we ever stopped to think that we human beings are just one of millions of species that have been created in this world? Does it mean that we have the sole right to dominate all the other species?

References IV

Aquinas, T. (1918) *Summa Theologica*, English translation by the English Dominican Fathers (New York: Benziger Brothers) Part II, Q. 64, Art. 1.

Aquinas, T. (1922) *Summa Theologica*, English translation by the English Dominican Fathers, 2nd revised edn (London: Burns, Oates and Washborne) Part I, Q. 93, Art. 4.

Aristotle (1985) *The Politics*, English translation by T.A. Sinclair, revised by T.J. Saunders (Harmondsworth: Penguin Books).

Baker, J.A. (1975) 'Biblical attitudes to nature', in H. Montefiore (ed.) *Man and Nature* (London: Collins).

Coon, C.S. (1962) *The History of Man* (Harmondsworth: Penguin).

Darwin, C. (1859) *On the Origin of Species by Means of Natural Selection or the Preservation of Favoured Races in the Struggle for Life* (London: John Murray; reprinted, Cambridge, Mass: Harvard University Press, 1975).

Darwin, C. (1899) *The Variation of Animals and Plants Under Domestication* 2nd edn (London: John Murray).

Davis, D.B. (1986) *Slavery and Human Progress* (New York and Oxford: Oxford University Press).

Davis, H. (1946) *Moral and Pastoral Theology* (London: Sheed and Ward) vol. II.

Dennis, T. (1986) 'Man beyond price: Gregory of Nyssa and Slavery', in A. Linzey and P.J. Wexler (eds) *Heaven and Earth: Essex Essays in Theology and Ethics* (Worthing, Sussex: Churchman Publishing).

Gardner, P. (1918) *Evolution in Christian Ethics*, Crown Theological Library (London: Williams and Northgate).

Goodin, R.E. (1983) 'Ethical principles for environmental protection', in R. Elliot and A. Gare (eds), *Environmental Philosophy* (Milton Keynes: Open University Press).

Hitler, A. (1974) *Mein Kampf*, English translation by D.C. Watt (London: Hutchinson).

Holcombe, W.H. (1986), cited in David Brion Davis, *Slavery and Human Progress* (New York and Oxford: Oxford University Press).

Huxley, A. (1973) *Brave New World* (Harmondsworth: Penguin).

IPPL Newsletter (1987), vol. 14, 2 July.

Jonas, H. (1984) *The Imperative of Responsibility: In Search of an Ethics for the Theological Age* (Chicago; London: University of Chicago Press).

Kant, I. (1909) *Fundamental Principles of the Metaphysic of Morals* 6th edn, translated by T.K. Abbott (London: Longmans).

Leopold, A. (1949) *A Sand County Almanac* (Oxford: Oxford University Press).

Mar Gregorios, P. (1987) *The Human Presence: Ecological Spirituality and the Age of the Spirit* (New York: Amity House).
McGourty, C. (1989) 'Abortion issue destroys board', Nature, vol. 341, p. 6.
Myers, N. (ed.) (1985) *Gaia Atlas of Planet Management* (London: Pan Books).
Nature (1988) BBC, 24 March.
National Anti-Vivisection Society (NAVS) (1987) *Biohazard* (London: NAVS).
National Anti-Vivisection Society (NAVS) (1988) 'Head transplant nightmare', *The Campaigner and Animals' Defender*, Nov/Dec, pp. 119–20.
Orwell, G. (1961) *Animal Farm* (Harmondsworth: Penguin Books).
Palmiter, R.D. *et al.* (1982) 'Dramatic growth of mice that develop from eggs microinjected with metallothionein-growth hormone fusion genes', *Nature*, vol. 300, pp. 611–5.
Palmiter, R.D. and Brinster, R.L. (1986) 'Germ-line transformation of mice', *Annual Review of Genetics*, vol. 20, pp. 465–99.
Paton, A. (1988) *Journey Continued* (Oxford: Oxford University Press).
Pope John Paul II (1988) *Solicitudo Rei Socialis*, Encyclical Letter (London: Catholic Truth Society).
Pursel, V.G. (1986), cited in *St Louis Post-Despatch*, Monday 8 December, p. 5.
Regan, T. (1983) *The Case for Animal Rights* (London: Routledge & Kegan Paul).
Ridley, M. (1988) 'Another slice from the tumour, dear?', *Meat Industry*, March, p. 10.
Runcie, R. (1988) 'Address at the Global Forum of Spiritual and Parliamentary Leaders on Human Survival', 11 April, pp. 13–4.
Ryder, R. (1983) *Victims of Science* (London: Davis-Poynter Ltd.).
Schank, R. (1988), cited in *Omni*, January; and also in *Agscene*, May, p. 3.
Singer, P. (1977) *Animal Liberation* (London: Paladin).
US PTO (1987) 'Device for perfusing an animal head', patent issued to Chet Fleming, 19 May, Patent No. 4,666,425.
Voltaire (1986) cited in David Brion Davis, *Slavery and Human Progress* (New York and Oxford: Oxford University Press).
Weld, T. (1986) cited in David Brion Davis, *Slavery and Human Progress* (New York and Oxford: Oxford University Press).
Wheale, P.R. and McNally, R.M. (1988) *Genetic Engineering: Catastrophe or Utopia?* (Brighton: Wheatsheaf Books, New York: St Martin's Press).

Part V
Public Acceptability and Control of Genetic Engineering

20 Introduction V

Dr PETER WHEALE and RUTH McNALLY

According to J.D. Bernal, science can be regarded as an institution, as a method, as a cumulative tradition of knowledge, as a major factor in the development of production, and as one of the most powerful influences moulding beliefs and attitudes to the universe and humanity (Bernal 1969). Indeed, science and technology have become so crucial to advanced industrial societies that they permeate the whole social system and have greatly increased the complexity of the problems which confront governments. One of the most important of such problems is how best to regulate and control scientific and technological activities.

Genetic engineering technology has enormous potential for improving welfare. But concern about the potential adverse environmental and animal welfare implications of genetic engineering and about the ethical and political implications of its application to analyse and manipulate human genetic material has reawakened the debate about its inherent dangers.

To what extent is genetic engineering acceptable to the general public? And how should this burgeoning technology be regulated? In the chapters that follow the contributors consider what can be regarded in a general sense as the 'governability' of genetic engineering.

Human Genome Research

The human genome is all the DNA which is characteristic of the human species. The EC, the USA and Japan each has a programme of research to map every human gene and determine the sequence of all the DNA in the human genome. Enthusiasm for human genome research is inspired by claims that it will provide the basis for the prevention or treatment of many human chronic diseases and will stimulate technological developments, for example in automated gene technologies.

Human genome research constitutes an example of what is sometimes referred to as 'Big Science' because of the vast sums of money it requires and the scale of the whole enterprise. One of the objectives of human genome research is the identification, isolation and appropriation of human genes which code for economically important proteins. There are believed to be over 50,000 different proteins in the human body (excluding antibodies) of which very few are currently used in medicine. Some of the genes which are discovered by human genome researchers may be patentable.

In Chapter 21 Edward Lee Rogers, legal adviser to Jeremy Rifkin of the Foundation on Economic Trends in the USA, explores some of the consequences of human genome research. Given the substantial rewards that will accrue to some companies as a direct result of human genome research, he calls into question whether it is fair that it should be financed by public funds. Rogers also argues that the application of genetic screening tests developed as a result of our increased knowledge of the human genome will raise a complex and interrelated group of ethical, socio-economic, legal and risk assessment questions. He is anxious to ensure that the individual's right to privacy is protected and that we do not see the emergence of a commercial eugenics movement. He reports that the Foundation on Economic Trends has proposed that human genome research in the USA should be controlled by a panel comprising representatives from all the groups that consider themselves to be vulnerable to discrimination because of the application of genetic screening tests developed from the knowledge gained from this research.

Public Opinion

Public opinion is an essential and integral component of liberal representative democracies and is fundamental to the political philosophy of the European Community (EC). In such democracies, the governments are elected by the people and public opinion influences government policies and legislation. It is therefore not unreasonable for government agencies and political parties to monitor the attitudes and perceptions of the general public in order to fashion proposed policies and legislation which are likely to be publicly acceptable.

In Chapter 22 Dr Joyce Tait, a Senior Lecturer in Environmental and Technology Management at the Open University, describes the data obtained from three public opinion polls which gathered information about public attitudes towards genetic engineering. These opinion polls were funded by the European Commission, the Department of Trade and Industry (DTI) in Great Britain and the Congressional Office of Technology Assessment (OTA) in the USA. Tait warns that the attitudes discerned through such opinion polls may not be a reliable guide to behaviour, and that a small but vocal minority can sometimes be effective in changing the opinion of the majority.

Public perceptions of science and technology may not, of course, be rational or based on full information. It is therefore important that well-balanced information is made available to the public in a form and in such a way as to facilitate its understanding of complex issues. However, according to the OTA opinion poll government agencies have a credibility problem with the general public. A two-thirds majority of the 1,273 US citizens interviewed in the OTA poll said they were less likely to believe information supplied by government agencies than that supplied by university scientists, public health officials and environmental groups (OTA 1987).

Harmonisation or Protectionism?

In recent times conscious efforts have been made towards the international harmonisation of the regulations controlling genetic engineering.

The Organisation for Economic Co-operation and Development (OECD), an alliance of 24 of the advanced industrial nations of the world, is a key organisation promoting this harmonisation process. In 1986 the OECD published a report entitled *Recombinant DNA Safety Considerations* (OECD 1986). This report concludes that there is no scientific basis for specific legislation for the implementation of genetic engineering techniques and applications and, like the 'Co-ordinated Framework for the Regulation of Biotechnology' in the USA, recommends that commercial applications of genetic engineering be regulated under existing product laws (OSTP 1986; Chapters 11 and 14).

Although the recommendations of the OECD report are not binding on its members, they are having a significant influence on the development of regulations and guidelines for genetic engineering worldwide. In Chapter 23 Brian Ager describes how the OECD recommendations have influenced the development of regulations in the UK and the European Commission's proposed draft directives on the contained use and deliberate release of genetically engineered organisms (European Commission 1988b, 1988c). Ager was the first Secretary for the Advisory Committee on Genetic Manipulation (ACGM) of the Health and Safety Executive (HSE) in the UK. In 1988 he was seconded to the OECD and to the European Commission's Concertation Unit for Biotechnology in Europe (CUBE).

The EC draft directives are part of the drive towards the harmonisation of regulations for genetic engineering activities throughout the EC and are consistent with the general integrationist policy which the European Commission is pursuing in line with the

Treaty of Rome. However, this policy is being directed towards the establishment of a single free market without the political integration which the original six Member States of the Common Market envisaged. This increase in economic integration without a parallel increase in the devolution of sovereignty to the European Parliament in order that it may exercise democratic control over the European Commission – the executive body of the EC – is a contradiction which could threaten liberal democracy in northern Europe. Indeed, we believe that the EC is facing a *constitutional* crisis comparable to that confronting the governments of Eastern Europe.

It can be argued that harmonised international legislation is necessary to control the deliberate release of genetically engineered organisms because these organisms can migrate or be transported from one country to another. Ager, for example, states that an international understanding of the safety issues in genetic engineering provides the basis for consensus on the protection of public health and the environment. He argues that it also leads to the promotion of technological and economic development and the reduction of national barriers to trade. However, whilst international harmonisation may reduce national barriers to trade within OECD countries, it may establish what are, in effect, non-tariff barriers to trade for the so-called Third World countries. One major problem with the present policy of international harmonisation is that those countries that are not advanced in genetic engineering technology, that do not have a science and technology policy in this field, have no intellectual property rights on biotechnological inventions, and that do not have a regulatory structure in place, will be the countries most vulnerable to exploitation and genetic expropriation.

The widening of the range of patentable inventions to include living organisms and their parts, as has occurred in the USA in recent years and is proposed in the EC (European Commission 1988d; see also Chapter 1), will create a situation in which patent law could be used to protect the industries of the advanced industrial nations by restricting imports from developing countries on the grounds that they infringe existing patents. The 1986 negotiations of the General Agreement on Tariffs and Trade (GATT) were aimed at developing a multilateral framework of principles, rules and disciplines for dealing with international trade in counterfeit goods, including agricultural produce. At these negotiations in 1986 it was agreed that new rules and disciplines will be introduced to facilitate the protection of intellectual property rights. However, Third World countries have little influence in GATT and it is unlikely that their interests will be favoured by these changes. As Calestous Juma observes, the harmonisation of the regulations in industrialised nations, and particularly those relating to intellectual property rights, will reinforce the process of what he calls 'genetic imperialism' (Juma 1989).

Pluralism and Technocracy

In pluralist societies, such as those of North America, Japan, Australia and New Zealand, and the countries of the EC, it is to be expected that sectional interests will seek to influence government. The advent of genetic engineering has prompted a range of interest groups to seek to influence government policy and regulations. These include, on the one side, animal welfare and environmental organisations, trade unions, religious and ethical organisations and small farmers' groups, and on the other side, industrial associations representing the interests of genetic engineering and pharmaceutical companies, chemical and agrichemical companies, and food and drink companies.

Roughly speaking, these two sets of groups have opposing views on the regulation and control of genetic engineering. Public pressure groups are campaigning for the introduction of measures which will retard, constrain or in some cases even prevent certain developments in genetic engineering, whilst the industrial interest groups are pressing for greater deregulation of genetic engineering and for a free market in the trade of the products and processes derived from it.

The drafting of legislation to cover genetic engineering requires that certain choices be made by the European Commission concerning the perception of potential risks arising from this technology and the acceptability of those risks. Whilst the leading industrial interest group – the European Biotechnology Coordinating Group (EBCG), a consortium of five major industrial organisations (see Poole *et al.* 1988) – was consulted during the drafting of the directive on the release of genetically engineered organisms into the environment, representatives of employees and representatives of environmental groups were not consulted. It is significant that the legal basis for the draft directives on the contained use and deliberate release of genetically engineered organisms is Article 100A of the Treaty of Rome which has as its objective the establishment of a single internal market for the EC by the end of 1992 rather than the protection of the environment, which would have been the case if the European Commission had chosen Article 130 of the Treaty of Rome. In the opinion of the Committee on the Environment, Public Health and Consumer

Protection of the European Parliament, the draft directive on the deliberate release of genetically engineered organisms into the environment is primarily a measure to regulate trade and is not aimed at establishing qualitative standards to protect the environment (Schmid 1989).

In Chapter 24 Benedikt Haerlin, who has been monitoring the development of EC regulations on genetic engineering for the European Parliament in his capacity as an MEP for the German Green Party, contends that industry has an unfair lobbying position because industrial pressure groups are well integrated into the daily administration and decision-making process of the European Commission. It is his view that the economic power of industrialists within the EC is succeeding in unduly influencing the Commission's draft directives to regulate genetically engineered organisms so that the emphasis is on the consolidated internal market of the EC and the relatively unimpeded development of genetic engineering technology, rather than on the establishment of qualitative standards for the protection of consumers, animals and the environment.

Haerlin argues that there is a need for a strong public counterpart to the power of the genetic engineering industry. However, the effectiveness of public pressure groups depends upon their ability to find scientists who are willing to co-operate with them by explaining technical issues. He believes that because scientists in academe must increasingly seek private funding for their research, few will risk compromising themselves with their patrons by providing the disinterested technical assistance which would enable the public or their representatives to enter into an effective dialogue with industry over the regulation and control of genetic engineering, and to participate on the official committees involved in the regulatory decision-making process, were they permitted to do so. The Athene Trust Conference participants raised the issue of the commercial pressures under which scientists labour which diminish their power to resist the sort of work which conflicts with ethical, environmental and animal welfare concerns. One suggestion, by Baroness di Pauli of the Institute of Social Invention, was that an ethical oath, equivalent to the medical profession's Hippocratic Oath, should be introduced for scientists and technicians (see 'Discussion V').

Social power can take many forms, from coercion and manipulation through to bargaining and persuasion. When social power is formalised within institutions, such as those of the EC, then the possibility arises that conflict will not be apparent. The predisposition of a social system can allow some interests to dominate others, creating a 'false consensus' in which resistance is not apparent because it is subsumed in the socio-economic structure and functioning of the social and economic system (see, for example, Partridge 1963; Wheale 1986; Wright 1986). This domination of certain interests – and its obverse, the subjugation of other interests – can be exercised through the economic structure, the composition and power of the bureaucracy, and a lack of democratic participation. Habermas (1971) has suggested that the politician is becoming the 'mere agent of a scientific intelligentsia' because technical decisions are made every day at a bureaucratic level by experts who are not held democratically accountable. When discussing the early Common Market, Lindberg and Scheingold wrote: 'It would not be an exaggeration to characterise the entire Community as essentially bureaucratic and technocratic' (Lindberg and Scheingold 1970).

The new techniques of genetic engineering constitute a technological revolution with the potential to alter our environment and our future. The purpose of the Athene Trust International Conference on genetic engineering was to provide a dialogue between scientists and regulators, and those interested in animal welfare, the environment and the ethics and morality of genetic engineering. A recurring theme, reflected in the papers presented and in the questions and statements made by conference participants, was that decisions regarding the appropriate course between the risks and benefits of genetic engineering should not be adjudicated solely within the inner circles of the scientific establishment. Participants generally agreed that, with disinterested advice from scientists, representatives of public interest groups can be effective in the regulatory process. There were repeated calls from the floor for provisions to be made for public participation in decision-making so that genetic engineering policy might embrace broader social and ethical issues as well as scientific curiosity, technical risks and commercial interests.

21 The Human Genome Project

EDWARD LEE ROGERS

The Human Genome Project is a massive US enterprise to map and sequence all the genetic information characteristics of human beings (*Mapping and Sequencing the Human Genome* 1988). Proponents of the Human Genome Project are demanding an annual budget of approximately US$ 200 million. The US Congress appears committed to fund this project although the actual level of funding for it is lower than its advocates would like because of the partially successful opposition to the project by pressure groups, including the Foundation on Economic Trends. For the fiscal year 1988, the US Department of Energy allocated US$ 10 million, and the NIH a little over US$ 17 million.

One of the justifications for federally funded human genome research is that it is claimed to be 'basic research' with the objective of isolating defective genes. Armed with such information, scientists may be able to develop therapies for genetic disorders. However, this would constitute 'applied research' which calls into question whether it should be financed by public funds or by private investment.

The Foundation on Economic Trends is concerned about who will control the human genome mapping and sequencing. Senate Bill 1966 proposes that a Gene Mapping Review Panel should be established which would be composed of government officials and representatives from industry and academe. The Foundation on Economic Trends has proposed an alternative panel that would have public representation from all the sub-groups that consider themselves to be vulnerable to the genetic screening procedures which are likely to result from the Human Genome Project. It is proposed that this alternative panel would have six members from the Senate, six members from the House of Representatives, and a committee appointed by the Human Genome Policy Board to include not only molecular biologists but also civil liberties experts and individuals who are knowledgeable and experienced in protecting the following: the health and safety of workers; genetic data; the rights of the handicapped; the rights of individuals undergoing antenatal and neonatal genetic screening; and the interests of consumers, especially in the field of medical insurance. The Foundation on Economic Trends has also recommended that the committee should also include experts in discrimination in education, bio-ethics and public affairs, and representatives of groups which have been victims of discrimination in the past, such as ethnic minorities, women and disabled people.

Genetic Screening

The mapping and sequencing of the human genome will provide new knowledge which may enable scientists to develop many new genetic screening tests to detect genetic defects, and to detect individual susceptibility to certain disorders. Such tests may be used to screen human foetuses, job applicants, employees, future spouses, and criminals, for example. The burgeoning genetic screening technology brings with it a complex and interrelated set of ethical, socio-economic, legal and risk assessment problems that cannot readily be resolved.

A major application of genetic screening tests is in prenatal diagnostics. Genetic screening tests are already used in conjunction with amniocentesis and chorionic villus sampling. Every year 50,000 amniocentesis tests are performed in the USA. If the screening tests performed after the amniocentesis indicate that a foetus has a genetic defect, the mother can choose to have an abortion. Until recently, the Supreme Court's 1973 decision in *Roe v. Wade* allowed a woman to choose whether or not to continue with a pregnancy for any reason during the first trimester (three months) of her pregnancy, and gave her limited rights of choice during the second trimester, under a constitutional claim of privacy. State law, therefore, could not impose any constraint on the exercise of those rights. The plurality opinion (four Justices) in *Webster v. Reproductive Health Services* (1989), appears, however, to give states the authority to impose restrictions on those rights.

The National Women's Health Network in the USA has suggested that amniocentesis and the testing of the foetus in the womb for chromosomal abnormalities denies women the right *not* to know the chromosomal status of the foetus they are carrying. The ability to know has evolved into the responsibility to know, and to act upon this knowledge, with the implication that a woman who knows her foetus has certain chromosomal abnormalities should abort it.

Genetic screening is a reality, and from a legal standpoint it is very difficult to ignore a technology that is available. If a gynaecologist were to fail to use a prenatal diagnostic test and subsequently a baby was born with a disability or a genetic disorder, he or she could be liable for malpractice. This liability would arise if a malpractice lawyer could prove that the gynaecologist knew of the diagnostic test, that it was generally available in the practitioner's community, and that it was standard procedure to use that diagnostic test.

It is believed that new screening tests developed as a result of the knowledge produced from human genome research will be able to detect and predict our predispositions to illnesses such as diabetes, heart attacks and liver disorders. We should distinguish between genetic screening tests for what we consider to be fatal or seriously disabling diseases which have an early age of onset, and tests for diseases which do not occur until later in life or which merely indicate an increased vulnerability to certain illnesses. Moreover there is a growing gap between our ability to diagnose an inherited genetic disorder, such as cystic fibrosis, sickle-cell anaemia, Lesch–Nyhan syndrome or Tay Sachs disease, and our ability to remedy that disorder (see also Wheale and McNally 1988a, Chapters 9 and 10; 1988b).

Job applicants and employees are also target groups for genetic screening tests designed to determine their susceptibility to certain disorders. If an unscreened employee with a genetic predisposition to a work-related illness were to become ill as a result of his or her work environment, it could be argued that the employer is liable for *not* screening him or her. This is one of the reasons why the insurance industry endorses pre-employment genetic screening.

The results of genetic screening testing could be useful to a prospective employer. Some argue that, as we already demand medical histories for insurance and employment applications, why should genetic information be treated differently? The reason is that genetic information on people broadens health information far beyond that of traditional health records. We all have some defective genes. I believe the results of genetic screening tests should be confidential and employers should be obliged to take their chances with the people they hire.

The use of genetic screening tests violates our right to privacy. Despite the Fourth Amendment of the US Constitution, which gives the right to be free from unreasonable searches and seizures, and the Fifth Amendment, which gives the right not to incriminate oneself, the US Supreme Court has held that the individual's right to privacy does not extend to preventing the government from collecting and revealing data about him or her.

The New Eugenics Movement

Genetic engineering techniques used in conjuction with the newly developed reproductive techniques, such as embryo transfer, may allow parents to avoid the birth of children with inherited genetic disorders and to choose to endow their children with certain physical characteristics thought desirable at the time (see Wheale and McNally 1988a, pp. 257–71).

Advocates of genetic screening argue that unwanted pain and suffering could be avoided by screening for specific disorders. They maintain that genetic screening is utility-orientated and benefits everyone by increasing our choices and decreasing our health risks. They do not believe that genetic screening will result in discrimination based on prejudice. But this argument is an over-simplification. Like the colour of our skin and our ethnic heritage, we have no choice or control over our genetic make-up. We have to determine who chooses whether or not the genetic testing is done, and the motive behind the testing – is it going to give medical benefits to the person screened or is it going to be more directly related to the enhancement or profitability of other parties? A new form of discrimination is going to emerge in the next decade, based not on skin colour or ethnicity, but on our genetic characteristics.

Genetic engineering, in combination with new human reproduction techniques such as artifical insemination (particularly in surrogate motherhood contracts) or embryo transfer, gives new significance and urgency to the eugenics debate. We should attempt to ensure that genetic privacy is protected, that constitutional rights are guaranteed and that we guard against the emergence of a commercial eugenics movement. In accord with basic human values we ought to be accepted as we are, as God made us, without our genetic vulnerabilities being measured, creating the possibility that they may be used against us.

22 NIMBY and NIABY: Public Perceptions of Biotechnology*

Dr JOYCE TAIT

The biotechnology industry is currently somewhat preoccupied with its rating in public opinion polls. This preoccupation is motivated partly by the fear that an unfavourable public perception of the industry could lead to obstructive public behaviour, depriving society of major new benefits or, at best, delaying their introduction and eating into the industry's profits. The recent experience of the chemical, nuclear and waste-disposal industries, where public opinion has proved irritatingly resistant to the influence of rational argument, heightens concern about the future public perception of biotechnology. Biotechnology is different in many respects from those industries currently suffering public relations problems. The word 'biological' still conjures up generally non-threatening images, and many biotechnology products will replace polluting chemicals or help to dispose of toxic wastes more safely. We will have new, more effective drugs with fewer side-effects. In addition, biotechnology is the first industry to be subjected to 'pro-active', rather than reactive, risk regulation.

An example of this in the UK is in the area of the deliberate release of genetically modified organisms into the environment for environmental or agricultural purposes. The Advisory Committee on Genetic Manipulation (ACGM) of the Health and Safety Executive (HSE) has developed its guidelines on this area in parallel with experimental developments, in particular with the programme of research on genetically engineered viral insecticides at the Institute of Virology (IoV) in Oxford (Bishop *et al.* 1988) [see Chapters 12 and 23]. The governments of the member states of the European Community (EC) are also taking a pro-active approach to the regulation of biotechnology, and there are moves to harmonise biotechnology regulation on an international basis (European Commission 1986) [see Chapters 20 and 23].

*A version of this chapter was first published in *International Industrial Biotechnology* (1988) vol. 8, no. 6, pp. 5–9, Cambridge University Press.

A pro-active regulatory approach is a major improvement on earlier approaches to risk regulation, where commercial products were released, or industrial wastes discharged, on an 'innocent until proved guilty' basis and, because proving guilt was sometimes a lengthy process, considerable damage was done to people or the environment by the time regulations came into force. However, as described below, pro-active risk regulation reinforces the need for more detailed information on public perception of the industry.

Some managers in the biotechnology industry and regulators in government bodies are optimistic about the future. They feel that the thoroughness of the regulatory initiatives for biotechnology is out of proportion to the real extent of the risks (assuming that we avoid the release of known pathogens). After all, they argue, we have been manipulating micro-organisms, plants and animals by conventional selective breeding for centuries, with no adverse effects [but see also Part I]; why should problems arise now just because we are using genetic engineering to speed up the process? The difference between the old methods and the new is quantitative rather than qualitative (Davis 1987).

Serious outbreaks of public resistance to new biotechnology developments, for example in the USA, West Germany and Denmark, should warn against undue complacency about public concern over the risks of biotechnology. The industry does have the potential to arouse strong passions, not least because it raises genuine moral and ethical issues for society. A small minority of the public, if sufficiently committed and vocal, can change the attitudes of the majority.

It is also possible that official assurances of safety will prove to be unjustified. Our past record in risk regulation contains many instances where we have failed to predict adverse impacts of new developments or failed to appreciate how ingenious (and dangerous) individuals and organisations might prove to be in exploiting a new technology.

This chapter summarises the analysis of the results of opinion polls on public perceptions of biotechnology, and discusses the factors that could exacerbate conflicts over future biotechnology developments. The relationship between our perceptions or attitudes and our behaviour is by no means straightforward (Ajzen and Fishbein 1980). Public opinion polls will not give sufficiently detailed information on the structure of people's attitudes and the relationships among them, or on the factors likely to influence their perceptions, to provide a basis for decision-making by industry, government or anyone else. Attempts to influence opinions or behaviour based on opinion poll data could have very unpredictable outcomes. A much more detailed investigation is needed to probe the structure of public attitudes, to understand their propensity to change, and to predict which attitudes are likely to be reflected in a person's behaviour and which will be suppressed or modified. Where conflicts do develop, a distinction should be made between those motivated by self-interest (NIMBY – 'not in my backyard') and those motivated by underlying values or ethical judgements (NIABY – 'not in *anybody's* backyard'). Failure to appreciate the importance of this distinction can seriously exacerbate conflicts over the risks from technology.

Opinion Polls

In this section I shall summarise the results of three opinion poll surveys which were conducted in the European Community (EC) as a whole, in Great Britain and in the USA.

In 1979 the Commission of the European Communities conducted a poll of public attitudes towards scientific and technological developments, including attitudes towards genetic research (see Cantley 1988). Some of the results of this survey are summarised in Table 22.1. The column headings on Table 22.1 relate to three different dimensions of attitude, so caution should be exercised in making comparisons across columns. However, there does seem to be a roughly reciprocal relationship between the extent to which genetic research is perceived to be worthwhile, and the extent to which it is perceived to present unacceptable risks. The position of West Germany and Denmark at the bottom of the list was perhaps an early indication of the problems of public acceptability of biotechnology that have arisen more recently in these countries. The UK and France were also positioned towards the bottom of this list, with a relatively large percentage of the population feeling that the risks were unacceptable but, despite the fact that both countries have made considerable progress with the field release of genetically engineered micro-organisms, no serious public dissent has yet arisen.

More recently the Office of Technology Assessment (OTA) has carried out a major survey of public perceptions of biotechnology in the USA (OTA 1987). This was based on 1,273 telephone interviews using a sample drawn from the civilian population aged 18 years and over. Households were selected by random digit dialing after the population had been stratified, based on four regions (east, south, midwest and west), and on type of place (central city, standard metropolitan statistical area (SMSA), and non-SMSA). After some preliminary questions about the benefits and risks of science and technology, the survey dealt with the understanding of genetic engineering, the uses of biotechnology in agriculture and medicine, and the regulation of its risks. The terms biotechnology and genetic engineering were used interchangeably in this survey.

Table 22.1:
Public attitudes to genetic research in the European Community in 1979

Country*	Genetic research felt to be:			
	Worthwhile	Of no particular interest	Unacceptable risks	Don't know
	(per cent)	(per cent)	(per cent)	(per cent)
Italy	49	19	22	10
Eire	41	20	22	17
Belgium	38	20	22	20
Luxembourg	37	31	18	14
Netherlands	36	17	41	6
UK	32	21	36	11
France	29	22	37	12
West Germany	22	16	45	17
Denmark	13	10	61	16
EC as a whole	33	19	35	13

Notes:
*Countries are listed in order of favourableness of public attitudes

Source: M. Cantley (1988) 'Biotech safety regulations and public attitudes in the EEC', *World Biotech Report 1988*, Proceedings of Biotech88 Conference, London, May 1988. Online Publications, pp. 77–88.

The report of this survey presents a picture of a public with mixed feelings, and sometimes with contradictory feelings, about biotechnology (Tait 1988). For example only one in five US citizens had 'heard about' any potential dangers from genetic engineering products, but 52 per cent believed that the products of genetic engineering were likely to present a danger to people or the environment. At the same time, 66 per cent thought that genetic engineering would make life better for all people. Among the most commonly cited hazards of biotechnology were (in declining order of frequency): difficulty in controlling growth or spread; health hazards, harmful effects; creation of mutations, monsters; environmental harm, contamination; unforeseen, unintended consequences; creation of new bacteria, diseases; causing cancer. The fact that only 12 per cent could cite a specific hazard of biotechnology was regarded as reassuring for the US government.

Table 22.2:
Public acceptability in the USA of risks from the environmental release of genetically engineered organisms

Risk Level*	Approve (per cent)	Not Approve (per cent)
Unknown	31	65
1:100	40	51
Unknown, but very remote**	45	46
1:1,000	55	37
1:10,000	65	27
1:100,000	71	21
1:1,000,000	74	18

Notes:
*Probability of a local ecological disruption from genetically engineered organisms

**These results indicate that the public acceptability of a risk level which is 'unknown but very remote' lies between their acceptance of a risk of 1:100 and 1:1,000

Source: OTA (1987) *New Developments in Biotechnology – Background Paper: Public Perceptions of Biotechnology* (Washington DC: US Government Printing Office OTA-BP-BA, May).

The OTA report states that: 'The public does not appear to be concerned about the morality of the genetic engineering of plants and animals' (OTA 1987, p. 57). This conclusion is based on the finding that 68 per cent of those surveyed stated that it was not morally wrong, and the 24 per cent who felt that it *was* morally wrong had a lower educational attainment or were more religious than the rest of the population. The report also states that among those who had moral objections to genetic engineering religious issues did not seem to be paramount, since only 31 per cent explained their objections in terms of religious beliefs or a belief in God, while 35 per cent objected on the grounds that 'people shouldn't tamper with nature'.

Provided there were no risks to people a majority of those surveyed approved of the use of genetic engineering for the following purposes: new treatments for cancer (96 per cent); new vaccines (91 per cent); cures for human genetic diseases (87 per cent); disease resistant crops (87 per cent); frost resistant crops (85 per cent); more productive farm animals (74 per cent); and larger game fish (66 per cent). The level of approval was thus highest for the uses perceived as being of the most immediate human benefit.

The OTA survey report also stated that the public was not averse to the risk of local ecological disruption from genetically engineered organisms that would significantly increase farm production with no direct risk to humans (see Table 22.2). A majority would approve at a risk level of a local ecological disruption from genetically engineered organisms of 1:1,000. The results of this survey indicate that the public acceptability of a risk level which is 'unknown but very remote' lies between their acceptance of a risk of 1:100 and 1:1,000.

In Great Britain a survey of public perceptions of biotechnology was carried out in 1988 by Research Surveys of Great Britain Ltd (RSGB) (1988) on behalf of the Department of Trade and Industry (DTI). This survey investigated the awareness, understanding and level of interest in biotechnology, and its perceived benefits and risks over the next ten years. Unlike the OTA survey (1987), genetic engineering was treated here as a separate issue, with questions asked about awareness, areas where research should continue and level of confidence in safety controls.

The questions on biotechnology and genetic engineering were included as part of an 'omnibus' survey of 1,917 men and women aged 16 and over. Respondents were selected for interview at 136 sampling points, spread nationally, via a random location sampling method. When the results were analysed, they were weighted to allow for minor sampling variations from the national population.

Only 38 per cent of the sample had heard of biotechnology compared with, for example, 91 per cent for silicon chips, 77 per cent for nuclear physics and 60 per cent for fibre optics. The only scientific term on the list which fewer people had heard of was superconductivity, with 23 per cent. This is similar to the results of the OTA survey where 35 per cent had heard or read 'a lot' or 'a fair amount' about genetic engineering, and 63 per cent 'relatively little' or 'almost nothing' (OTA 1987).

In answer to an open-ended question about the meaning of biotechnology, those who were aware of the term were generally unable to give more than a very vague answer, with medical or health-related applications predominating.

All interviewees were asked to respond to the list of applications given in Table 22.3, indicating which they would or would not associate with biotechnology. The bases given in Table 22.3 have been weighted to allow for sampling variations, as mentioned above.

The application most frequently associated with biotechnology was the production of drugs such as vaccines or insulin and antibiotics. As Table 22.3 indicates, those who were previously aware of biotechnology had a stronger tendency to associate *all* the applications with it, including those not actually associated with it. Those who were more highly educated were more likely to perceive all the applications as being associated with biotechnology, once again including those that were not relevant. The survey report notes that one of the most commonplace applications of biotechnology

Table 22.3:
Public awareness in Great Britain of practical applications associated with biotechnology^ (Base: 2,000 adults)

(Base)	Previously aware of biotechnology		Educational qualifications		
	Yes	No	None	A-level & below	HND & above
	(757)	(1243)	(1041)	(613)	(182)
Vaccines	65	45	45	59	67
Production of insulin/antibiotics	60	37	38	52	68
Organ transplants*	54	38	37	51	57
Fertility drugs*	55	33	33	50	59
Finding a cure for AIDS	51	35	34	47	56
Disease resistant crops	55	33	34	46	66
Human embryo transplants*	54	32	33	47	52
Cloning of animals/crops	58	29	31	44	72
Genetic engineering	58	27	28	46	70
Producing new foods/crops	50	31	32	40	65
Biological washing powders	42	34	37	33	51
Breakdown of oil slicks/waste	41	26	27	32	55
Natural food flavourings	35	23	25	28	38
Artificial limbs*	30	21	21	29	32
Making cheese	30	12	13	21	38

Notes:
^Figures are percentage of respondents
*Application *not* associated with biotechnology

Source: Research Surveys of Great Britain Ltd (1988) *Public Perceptions of Biotechnology: Interpretative Report* (RSGB Ref. 4780 March).

was the one least thought to be associated with the subject – making cheese.

Table 22.4 summarises responses to the question of whether biotechnology will improve or harm the quality of life over the next ten years. This question was asked on three occasions during the survey as interviewees gradually built up an understanding of what was involved. The more information people were given the more they seemed to feel that biotechnology would improve the quality of life, but there was a slight trend in the reverse direction for those who thought it would improve it in some ways and harm it in others. It is important to note that people were only given information about the potential benefits of biotechnology; provision of information about its hazards would probably have had the reverse effect.

Among the 13 per cent of respondents who said that biotech-

Table 22.4:
Opinions in Great Britain on whether biotechnology will improve or harm the quality of life over the next ten years

	Improve	Harm	Improve in some ways, harm in others
	(per cent)	(per cent)	(per cent)
Aware of 'biotechnology'*			
(1)	42	9	-
(2)	63	3	12
(3)	66	4	15
Unaware of 'biotechnology'*			
(1)	2	1	-
(2)	46	4	5
(3)	56	3	7

Notes:
*Questions about the impact of biotechnology on the quality of life over the next ten years were asked at three different times during the questionnaire: (1) at the beginning, in the context of six other scientific terms such as silicon chips and nuclear physics; (2) after the presentation of the 15 possible applications of biotechnology, listed in Table 22.3; (3) after the presentation of the same list of biotechnology applications, excluding the four that are not associated with biotechnology.

Figures given are percentages of the total sample in each category; the difference to 100 per cent is made up largely of people who did not know plus a small number (5 per cent or less) who thought there would be no effect.

Source: Research Surveys of Great Britain Ltd (1988) *Public Perceptions of Biotechnology: Interpretative Report* (RSGB Ref. 4780 March)

nology would harm the quality of life in some way, the applications involved included: cloning of animals and crops (63 per cent), genetic engineering (50 per cent), producing new foods/crops (27 per cent), natural food flavourings (20 per cent), disease resistant crops (17 per cent), and biological washing powders (12 per cent).

Respondents were asked to say whether genetic engineering research should continue in six possible areas, involving direct or indirect effects on humans. Where direct effects were implicated, the production of drugs to treat chronic diseases such as cancer and heart disease from altered micro-organisms was rated most highly (77 per cent), followed closely by the alteration of human cells to prevent inherited diseases. There was little support (13 per cent) for research to improve the level of intelligence that children would inherit. (The inclusion of this last item seemed strange, given that it is not currently a feasible application of genetic engineering.)

Where indirect effects on humans were involved, there was relatively little support for changing the make-up of animal cells to breed more productive farm animals (24 per cent); 41 per cent felt that experiments in which micro-organisms are released to control agricultural pests should continue, and 68 per cent approved of altering plant cells to produce disease resistant crops.

These results indicate that there is concern about experimentation with animal cells and, to a lesser extent, about releasing genetically engineered micro-organisms, with genetic engineering involving human or animal cells for non-life-saving reasons being most unacceptable. Medical applications were the most acceptable.

Confidence in the adequacy of safety controls was fairly evenly divided. For genetic engineering, 40 per cent were very confident or quite confident and 41 per cent were not very confident or not at all confident; similar results were found for biotechnology. Those who had less confidence in the control of biotechnology were less likely to see it as improving the quality of life, more likely to see it as harmful, and less likely to think that the research should continue.

Limitations of Opinion Polls

Opinion polls about risk perceptions obviously provide information that governments and companies value, otherwise they would not continue to be funded. Opinion polls indicate to an industry whether it already has a serious public image problem, whether different sectors of the population have different perceptions and, if carried out regularly, whether perceptions are changing with time. However, there are pitfalls.

There are often inadequacies in the interpretation of data from opinion polls and they cannot provide much of the information on risk perceptions that is needed to support a pro-active approach to risk regulation. Another pitfall with opinion polls occurs particularly with a topic such as biotechnology, where the public does not yet have a clear understanding of what it means or a well-formulated set of opinions about it. In such circumstances minor changes in the wording of questions can result in apparently large swings of opinion or, as shown by the UK example above, information provided in the course of a survey can lead to major, but probably unstable, changes of perception.

There is also a tendency, which was particularly apparent in the OTA survey quoted above, to adopt a vote-counting approach to the interpretation of the data – as long as those in favour of biotechnology are in a majority, the views of the minority can be ignored. This overlooks the fact that a vocal and forceful minority can often bring about major changes in public opinion or can induce active dissent in a silent majority.

Opinion polling is inevitably a 'broad-brush' technique. It does not provide *detailed* information on the nature of perceptions about an issue, why they are favourable or adverse, whether these perceptions will remain latent or are likely to be translated into actions, and for which groups this is most likely to be the case. Most importantly, it does not indicate the extent to which any potential conflict is motivated by concern for the interests of protagonists or by ethical and value-based considerations. The former is epitomised by the well-known NIMBY syndrome ('not in my backyard') (Gervers 1987), but the distinction between this and the alternative NIABY syndrome ('not in anybody's backyard') is rarely recognised. NIMBY tends to be used as a blanket term to describe all conflicts over potentially hazardous or otherwise unwanted activities.

Some of the differences between NIMBY and NIABY issues are indicated in Table 22.5. In any community of individuals who are concerned about an issue, for some it will be NIMBY, for some NIABY, and for others a mixture of the two; a continuum of shades of attitude will be held both by individuals and between individuals.

There is a need to understand the extent to which an issue can be characterised as NIMBY or NIABY, and the extent to which the attitude is likely to change from one to the other. A different approach from opinion polling is needed to understand attitudes or perceptions at this fine level of resolution. Appropriate techniques are being developed at the frontiers of research in several areas of social science (Eden *et al.* 1983; Tait 1983; Nelkin 1985) but lack of government funding in recent years, particularly in the UK, is affecting progress seriously. Common themes in this methodological research are: the need to find ways of enabling those surveyed to express their perceptions in their own terms, rather than subjecting them to the artificiality of a standard questionnaire, and the need to analyse the resulting data so as to enable valid generalisations and predictions to be drawn from it.

Table 22.5:
The differences between NIMBY and NIABY conflicts over potentially hazardous activities

NIMBY	NIABY
Based on self-interest of protagonists	Based on ethics or values of protagonists
Likely to be restricted to specific biotechnology developments	Likely to spread across all biotechnology developments
Likely to be location-specific	Likely to be organised nationally or internationally
Can usually be resolved by the provision of information or compensation, or by negotiation	Very difficult to resolve – information is viewed as propaganda, negotiation as betrayal, compensation as bribery

Notes:
NIMBY   Not in my backyard
NIABY   Not in anybody's backyard

In industries with reactive risk regulation, for example pesticides or nuclear power, opinion polling approaches are comparatively straightforward. In such cases there is an identifiable population with, probably, strongly-held and stable opinions about the issue. Opinion polling about industries subjected to pro-active risk regulation, as is being attempted for biotechnology, is more difficult. As indicated by the opinion polls described above, most respondents have only a vague idea about the issue, have never seriously thought about it until presented with the questionnaire, and therefore have labile and poorly formulated opinions. It is important to identify which individuals or groups are likely to be influential in the future, and the direction of their influence.

Where biotechnology is concerned, it could be predicted tentatively that animal welfare groups and various 'pro-life' medical pressure groups are likely to become involved in influencing public opinion because their current spheres of concern already overlap with some key areas of biotechnology development. Such pressure groups will probably attempt to present biotechnology as a NIABY rather than a NIMBY issue, partly because the philosophy of these groups is based on a shared system of values rather than of interests, and partly because it is easier to mobilise a broad spectrum of public opinion around a value-based issue than it is about an interest-based issue. As summarised in Table 22.5, this would tend to spread conflicts over biotechnology to a wider range of developments over a wider geographical area, and also make them more difficult to resolve.

Another important influence on the future of the biotechnology industry has yet to be discussed. This is the behaviour of regulators and of managers in the industry itself. The people employed in both these areas are subject to social processes in much the same way as is the rest of the population. Their choice of actions and their choice of words to express their views will be a major determinant of the public's perception of biotechnology. There is, therefore, an equally urgent need to study the perceptions of industrialists and regulators to find out how they perceive the biotechnology industries they are involved with, and also how they perceive the public as potential arbiters of these industries.

The Need for Research and Dialogue

The aim of research on attitudes and values relevant to the biotechnology industry should be to improve the understanding among industrialists, regulators and the public of the issues surrounding new biotechnology developments. Such research also enables industrialists, regulators and the public to become cognisant of each other's views. One possible outcome of successful research in this area would be the prevention of *unnecessary* escalation of conflict of the NIABY type. However, it should not be assumed that removal of all areas of conflict is desirable; some conflict can be productive and can lead to more effective regulation than would otherwise be the case, and for some people there will always be ethical objections to some aspects of biotechnology.

Research on public perceptions of genetic engineering should help in understanding and responding to the needs and requirements of the public. It should not seek to manipulate public opinion according to the needs of regulators or industry. Any hint of the latter would merely add the stigma of social engineering to the image of genetic engineering.

One way of ensuring effective dialogue between industry, regulators and the public would be to create a forum where they can discuss issues openly and on an equal basis, in the expectation that relevant action will stem from the discussions. Biotechnology is breaking new ground in attempting a pro-active approach to risk regulation. The type of discussion forum advocated in this chapter, backed up by adequate research information, would have been valuable for much recent reactive risk regulation, but for pro-active regulation it is even more valuable. In its absence industry and regulators will be feeling their way in the dark, and the benefits in terms of more effective regulation, better public relations and a smoother transition to product innovations may slip through their fingers.

23 The Regulation of Genetic Manipulation

BRIAN AGER

Historical Background

Genetic manipulation is the name used by the regulatory authorities in the UK for the techniques developed in the early 1970s that make it possible to cut and splice genes and to move them from one species to another, regardless of their sexual compatibility or their evolutionary relationship. The safety aspects of genetic manipulation have been the subject of considerable debate. Early fears included the possibility of turning relatively harmless and commonplace micro-organisms into dangerous pathogens or into cancer-causing agents. More precisely, these fears stemmed from the fact that the majority of genetic manipulation work uses the bacterium *Escherichia coli (E. coli)*, a normal inhabitant of the human gut. One of the early safety questions raised was whether it would be possible through genetic manipulation inadvertently to turn this relatively harmless organism into a pathogen (disease-causing organism) which could then displace its natural counterparts from the human gut. Also, were a laboratory worker to become colonised by *E. coli* into which a gene coding for a toxin (biological poison) had been introduced, could the toxin be produced at a site within his or her body? These concerns prompted scientists in the USA to agree to a self-imposed moratorium on certain genetic manipulation experiments.

By the end of the 1970s these early fears had come to be widely regarded as exaggerated. Not only had genetic manipulation developed an excellent occupational health and safety record, which it continues to maintain, but increased knowledge of molecular biology laid to rest some of the earlier fears. Also, there had been considerable progress in the development of 'disabled' host bacteria and vector systems for genetic manipulation work. These 'disabled' bacteria are strains that are limited in their ability to survive outside the laboratory environment.

A further important feature for the safe development of genetic manipulation was the establishment of national advisory committees. In the USA, the Recombinant DNA Advisory Committee (RAC) was set up in 1975 and in the UK the Genetic Manipulation Advisory Group (GMAG) was established in 1976. (The GMAG was disbanded in 1984 and replaced by the Advisory Committee on Genetic Manipulation (ACGM).) The RAC and the GMAG and their supporting government agencies were foremost in developing a balanced regulatory framework for genetic manipulation. The GMAG, in particular, was constituted with a broad membership comprising representatives of industry and trades unions together with scientists and representatives of public interest groups. These national advisory bodies were reponsible for the institution of detailed guidelines to control work with genetically manipulated organisms. These guidelines were generated in parallel with the development of the technology rather than as a consequence of any occupational disease or other adverse effect. Towards the end of the 1970s, as knowledge increased and the hazards remained conjectural, so the guidelines were relaxed in the USA, the UK and worldwide.

From a health and safety perspective genetic manipulation is an unusual activity. With other activities, including work with pathogens, controls have often been developed as a result of a demonstrated incidence of ill-health, injury or mortality. Although with genetic manipulation work the hazards have so far proved to be conjectural, from the outset it has been the subject of a detailed regulatory structure in the UK. This regulatory structure is described below together with some international developments towards harmonisation in this area.

The UK Regulatory Structure

In this section on the UK regulatory structure I identify seven elements which are relevant to genetic manipulation activities.

1. Legislation – The Health and Safety at Work Act (1974)

The basis for regulatory control of genetic manipulation in the UK is the Health and Safety at Work Act (1974) (HSW Act). This Act places a general duty on the employer to provide and maintain a safe working environment for employees. Employers are also charged with a duty to avoid exposing to risks those not in their employment, including the general public. These general duties are qualified by the phrase 'so far as is reasonably practicable'. This means that the employers must make a cost-risk analysis and assess on the one hand the risk of the work and, on the other hand, the difficulty and expense involved in avoiding that risk. Greater risks require greater precaution.

Genetic manipulation and biotechnology fit comfortably into the framework of the HSW Act, and the general duties of the Act apply to this area as they do to any other work activity.

2. *Notification – The Health and Safety (Genetic Manipulation) Regulations*

In August 1978 the Health and Safety Commission (HSC) issued the Health and Safety (Genetic Manipulation) Regulations (1978), under the HSW Act. These regulations made it legally binding for any centre intending to undertake any activity involving genetic manipulation to give prior notification to the inspectorate of the HSC – the Health and Safety Executive (HSE). This notification involves certain details about the facilities, the on-site arrangements for assessing risks (see element 5 below), details about the monitoring of worker health on site, as well as certain details of individual experiments. Prior notification is required for experiments assessed as being in the high risk categories. Experiments in the low risk categories are notified to the HSE retrospectively each year.

The Health and Safety (Genetic Manipulation) Regulations (1978) did not apply to large scale work (regarded as fermentation in volumes of 10 litres and above) with genetically manipulated micro-organisms, or to the planned release of genetically manipulated organisms into the environment. Such activities were only covered by non-mandatory notification schemes. However, proposals for new Genetic Manipulation Regulations were set out in a consultative document by the HSC in September 1987 and new regulations were issued in 1989. The main changes are the extension of mandatory notification to planned release and to large scale work. The new regulations also include a reduction in notification detail for low-risk work and a legal obligation to establish local genetic manipulation safety committees (see element 5 below) are also included.

Figure 23.1 illustrates how the notification requirements of the regulations operate. There are several advantages in having this type of notification requirement. Mandatory notification allows the HSE, the ACGM and government departments, where relevant, to review local arrangements for overseeing genetic manipulation work from the start. It identifies those organisations who should receive the detailed guidance on genetic manipulation activities produced by the ACGM (see element 4 below) and it enables the HSE and the ACGM to review high-risk work before it takes place (discussed below). Most importantly, it allows the HSE to target inspection which is carried out by specialist inspectors from the HSE's Technology Division. These inspectors are responsible for the inspection and enforcement of HSW Act duties in all genetic

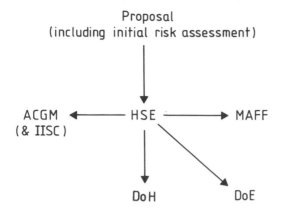

Notes:
HSE   Health and Safety Executive
AGCM   Advisory Committee on Genetic Manipulation
IISC   Intentional Introduction Sub-Committee (formerly Planned Release
         Sub-Committee)
DoH   Department of Health
DoE   Department of the Environment
MAFF   Ministry of Agriculture, Fisheries & Food

Figure 23.1:
Schema of Advisory Committee on Genetic Manipulation (AGCM) notification (1988)

manipulation centres including field sites, laboratories and large scale fermentation sites.

3. Advice – Advisory Committee on Genetic Manipulation (ACGM)
Proposals to undertake genetic manipulation are considered by a specialist committee, the Advisory Committee on Genetic Manipulation (ACGM). This is the watchdog committee that was set up in March 1984 to replace the former GMAG, which was active from 1976 until 1984 and was operated under the aegis of the Department of Education and Science. Unlike GMAG, the ACGM operates under the HSC and the HSE and not directly under any government department. The HSE's Medical Division provides the Secretariat to the ACGM.

The ACGM has a tripartite structure consisting of five representatives of employers, five employee representatives and eight scientific specialists. It has an independent Chairperson, presently (1989) Sir Hans Kornberg, Professor of Biochemistry, Cambridge

University. Three of the employer representatives are from the Confederation of British Industry (CBI), one is from the Research Councils, and the other is from the Committee for Vice Chancellors and Principals. The employee representatives are all nominated by the Trades Union Congress (TUC). Amongst the eight scientific specialists there is a local authorities' association nominee.

One of the roles of the ACGM is to advise the HSC and the HSE in connection with their duties under the HSW Act. But the ACGM is also an 'open shop' to other government departments. These include the Ministry of Agriculture, Fisheries and Food (MAFF), the Department of the Environment (DoE), the Department of Trade and Industry (DTI) and the Department of Health (DoH). The ACGM is able to advise these government departments on safety, scientific and technical issues that may arise in connection with their duties and their legislation.

4. Guidance – ACGM Guidelines

The fourth element is the programme of work of the ACGM from 1984 onwards. It is important that the scientific basis which supports the guidelines and on which proposals are judged is open to review as further experience is gained and the results of research accumulate. The ACGM has been extremely active in revising the guidelines issued by the GMAG. For example, in September 1988 it issued a revision of the GMAG risk assessment guidelines (see Box 23.1) and laboratory containment categories. The ACGM also produces guidelines on new topics. Planned release of genetically manipulated organisms into the environment (see below) and laboratory work with oncogenes (cancer genes) were selected as the first two priority topics requiring specific guidelines when the ACGM was set up in 1984. The ACGM has issued guidelines on these topics and on viral vectors, large scale use of genetically manipulated organisms (see below), the health surveillance of workers in the laboratory and in large scale production facilities, transgenic animals (see Box 23.2) and on disabled host-vector systems.

The ACGM develops its guidelines by setting up specialist working groups. These working groups are responsible to the ACGM, and are chaired by a member of the ACGM, but may have an even broader membership depending on the topic. Each working group produces draft guidelines and reports back to the main committee of the ACGM. Once the main committee is satisfied with the draft guidelines they are presented to the HSC for its endorsement before they are issued to genetic manipulation centres and to other interested bodies. This process demonstrates

---

Box 23.1:
ACGM risk assessment scheme for laboratory work

Work with pathogens is the closest relevant activity to laboratory genetic manipulation work. In work with pathogens, risk assessment, and hence the selection of appropriate physical containment measures, is based on a consensus of experts. The Advisory Committee on Dangerous Pathogens (ACDP) has published detailed guidance categorising pathogens into hazard groups (ACDP 1984). For a given pathogen, the type of questions pertinent to risk assessment are, 'What is the severity of disease produced?', 'What is the infectious dose (if known)?', 'Is treatment available?', and 'Is the pathogen capable of causing epidemics?' In other words, 'punishment' on the basis of a known 'criminal' record and, in this case, 'trial' by the ACDP. For genetic manipulation, however, there is no such record of harm on which to base controls. What was developed in the UK instead was a ranking system, or semi-quantitative risk assessment scheme, based on considering each experiment under the headings 'Access', 'Expression' and 'Damage' (GMAG Note 14).

Access is the chance that the manipulated organism will be able to colonise the human body. *Expression* is the expected or known level of expression of the inserted gene. *Damage* is the chance of the gene product giving rise to an ill-health effect in a worker exposed to a manipulated organism.

Each experiment is assigned a factor [a numerical value] under each of the three headings. The factors are then multiplied to give an overall score which in turn dictates the level of containment. This system forms the basis of risk assessment for laboratory genetic manipulation experiments in Great Britain. The ACGM issued new guidance based on this scheme in 1988 (ACGM/HSE/Note 7).

---

how the ACGM is constantly seeking to address new areas as well as rethinking existing guidance in the light of experience.

5. Local Participation – Local Genetic Manipulation Safety Committees
The fifth element in the UK regulatory structure is local participation. Under the genetic manipulation guidelines each genetic manipulation centre must appoint what is called the 'Local Genetic Manipulation Safety Committee'.

The constitution of such local committees is a key element. There has to be a balance between those representing management and those representing employees. A typical Local Safety

Box 23.2:
The regulation of the transgenic manipulation of animals in Great Britain

The Advisory Committee on Genetic Manipulation's (ACGM's) Planned Release Sub-Committee set up a working party to review the regulation of transgenic manipulation of both laboratory and agricultural species. This working party, chaired by the ecologist Dr Mark Williamson, Professor of Biology at the University of York, comprised representatives from the Home Office, the Department of the Environment (DoE), the Ministry of Agriculture, Fisheries and Foods (MAFF), the Nature Conservancy Council (NCC), the Agriculture and Food Research Council (AFRC) and the Health and Safety Executive (HSE). It reported to the ACGM through the Planned Release Sub-Committee and the *Guidelines on Work with Transgenic Animals* were endorsed by the Planned Release Sub-Committee and the ACGM, and published in January 1989 (ACGM/HSE/Note 9).

The transgenic animals guidelines cover three main areas, namely, the health and safety of workers who develop transgenic animals; animal welfare issues, and environmental issues. Detailed guidance is given on risk assessment, the need for containment, notification requirements and what constitutes environmental release in the context of transgenic animals.

Welfare aspects are also covered under existing Home Office and MAFF arrangements. The Home Office makes it clear that no animal will be released from the control of the Animals (Scientific Procedures) Act (1986) until it is clear that no harm to the animal will result. This act, which is administered by the Home Office, regulates, 'any experimental or other scientific procedure applied to a protected animal which may have the effect of causing that animal pain, suffering, distress or lasting harm.' A protected animal is any living vertebrate other than humans.

Section 2 (3) of the act is of prime importance in the context of transgenic animals. This section states that, 'anything done for the purpose of, or liable to result in, the birth or hatching of a protected animal is also a Regulated Procedure if it may cause pain, suffering, distress or lasting harm'. There are two aspects of the production of transgenic animals that require consideration in the context of this act, namely, the insertion of DNA into the germline, and the subsequent breeding of animals carrying the desired characteristic. Transgenic animals are produced by germ-line manipulation [see Chapter 3]. The procedures to which the

---

Box 23.2 continued ...

donors and recipients in germline manipulation are subjected are Regulated Procedures within the meaning of the 1986 Act. The initial production of transgenic animals is therefore regarded as a Regulated Procedure and should be carried out under the authority of a Project Licence and a Personal Licence. Breeding from transgenic animals is also regarded as a Regulated Procedure requiring Project and Personal Licence authority until it can be demonstrated that the progeny are not likely to suffer adverse effects.

Committee comprises a Chair, a Biological Safety Officer, a Supervisory Medical Officer and representatives chosen by and from all those persons who have access to the laboratory facility. The correct make-up of these committees is insisted upon by the ACGM and the HSE during initial notification.

These local committees have a number of functions. They review in detail each proposed project and undertake an initial risk assessment before notification is sent to the HSE. They assess whether staff training and experience are sufficient to carry out the work safely and whether the arrangements for health monitoring are adequate. They also review laboratory practice and consider any accidents or dangerous occurrences that might occur in their facility.

6. Inspection – Health and Safety Executive (HSE)
The HSW Act has a clear emphasis on self-regulation in that the employer is required to assess risks that may arise and to take appropriate precautions. Nonetheless, powers exist to enforce the duty required by this act. The powers are the same whether we are considering the construction industry, the nuclear industry, work with pathogens or genetic manipulation.

On-site inspection is an important function of the HSE. It underpins the whole regulatory system and is conducted on a priority basis by HSE inspectors whose function it is to enforce the general duties of the HSW Act. HSE's Technical Division includes a group of inspectors specialising in the oversight of genetic manipulation activities. Not only do HSE inspectors consider the condition of on-site facilities and equipment, they carefully consider factors such as local codes of practice, the administrative management of safety issues, and whether the experimental procedures are being carried out safely, taking national guidelines into account.

The above six elements of the regulatory structure for genetic manipulation in Great Britain comprise a detailed control package which was put in place soon after the technology was developed, and then adjusted in the light of experience and new knowledge. There are few other examples of this type of pro-active regulation. Once the safety concerns had been expressed, it was thought necessary to implement a system of risk assessment and of risk management. In order to inspire confidence, particularly that of those outside the scientific community, this system had to provide broadly based scientific advice, supported by the law. Given the conjectural nature of the hazard an important feature of the system was, and still is, its flexibility, which permits the guidelines and procedures to be adjusted with experience.

7. Product Regulation
In addition to the previous six elements of regulation which are specific to genetic manipulation work there is a seventh element: the set of product legislation that would apply when a product of genetic manipulation is brought to the market. Examples of relevant product legislation include pesticide regulations, regulations for medicines – be they human or veterinary – and food regulations.

International Harmonisation

An important step towards international harmonisation in this area has been made by the Organisation for Economic Co-operation and Development (OECD) with the publication of a Report entitled *Recombinant DNA Safety Considerations* (OECD 1986). The OECD is an inter-governmental organisation comprising 24 countries – 19 European countries, Australia, Canada, Japan, New Zealand and the USA. The report is the work of an *ad hoc* group of government experts established by the OECD's Committee for Scientific and Technological Policy. Its recommendations were adopted by the Council of the OECD, where countries are represented at their highest level. The use of the term 'recombinant DNA (rDNA) organism' in the report is broadly equivalent to the term 'genetically manipulated organism' used by the regulatory authorities in the UK. The central task of the ad hoc group was to identify scientific criteria for the safe use of rDNA organisms in industry and in the environment, rather than the construction of such organisms in the research laboratory.

The main conclusions of the OECD report are as follows. Any risks raised by the use of rDNA organisms are expected to be of the same nature as those associated with conventional organisms. Furthermore, such risks may be assessed in generally the same way as for non-rDNA organisms. Although rDNA techniques may result in organisms with a combination of traits not observed in nature, the properties of such organisms will be inherently more predictable than organisms genetically modified using conventional methods. The report also concludes that there is no scientific basis to justify *specific* legislation for the use of rDNA organisms – although it recommends that existing regulations be examined to ensure that such uses are adequately controlled.

The European Commission has also become increasingly interested in the possibility of community-wide biotechnology regulation, and towards the end of 1986 set out its intentions in a Communication to Council for *A Community Framework for the Regulation of Biotechnology* (European Commission 1986). This Communication notes that the techniques of genetic manipulation are finding increasing application, and that several countries are reviewing existing regulations taking account of the OECD report (OECD 1986). The Commission described its intentions to introduce proposals addressing two distinct aspects of the use of genetic engineering, namely authorisation of planned release of genetically engineered organisms into the environment, and levels of physical and biological containment, accident control and waste management in industrial and laboratory applications.

The European Commission subsequently drafted three proposals for Council Directives relevant in this area (European Commission 1988a; 1988b; 1988c). The Commission's draft *Proposal for a Council Directive on the Protection of Workers from the Risk Related to Exposure to Biological Agents at Work* (European Commission 1988a) aims to protect workers from the risks related to exposure to biological agents able to cause human disease, whether genetically modified or naturally occurring. The draft directive is divided into two main parts: a general one applying to all sectors of activities where workers may accidentally be exposed, and a second part applying only to those whose work involves the handling of pathogens. Activities involving genetic manipulation included in the second part require advance notification when they involve biological agents pathogenic to humans. Accidental releases which are likely to involve a danger to the health of workers also require notification. A list of containment principles are envisaged for healthcare facilities, diagnostic laboratories, industrial processes, laboratories and animal rooms. The other two draft directives, on the large scale contained use and deliberate release of genetically engineered organisms, are discussed below.

There is little doubt that the OECD report (1986) has had, and will continue to have, considerable influence on the development of regulations and guidelines worldwide. It has influenced the regulation of genetic manipulation in many countries including the UK, Australia, the Netherlands, the USA, Japan, West Germany and France, and the proposals of the European Commission. In the UK the influence of the OECD report can be seen in the guidelines for planned release (ACGM/HSE/Note 3) and the guidelines for large scale work (ACGM/HSE/Note 6). Below I shall examine its influence on both the UK regulations and the EC proposed directives on large scale contained use and planned release.

Large Scale Contained Use
The principal recommendations in the OECD report on the large scale contained use of rDNA organisms are as follows. The vast majority of cases of industrial large scale applications to date have used rDNA organisms of intrinsically low risk. The report identifies criteria for classifying organisms as low risk, and recommends a corresponding level of control based on established good industrial practice (Good Industrial Large-Scale Practice [GILSP]) for their use. The report lists the basic principles of GILSP, which should underpin all large scale applications of rDNA organisms whatever the risk. For rDNA organisms of higher risk categories it recommends additional control and containment options.

When drawing up their guidelines, the ACGM takes into account international activities. As a consequence the regulatory approach being taken in the UK is broadly in line with that being adopted in several other countries.

For large scale work, widely regarded as work with volumes in excess of 10 litres, there are separate ACGM guidelines (ACGM/HSE/Note 6). These large scale guidelines make the point that although there are no known health hazards specific to genetic manipulation, complications of hazards in traditional biotechnology (for example, enhanced immune response to human proteins expressed in combination with bacterial proteins) can be envisaged. Nevertheless the ACGM's view is that there is no qualitative difference between the hazards of using genetically manipulated organisms on a large scale, and constructing them in the laboratory. Scaling up increases the magnitude of operation and hence the potential for worker exposure to genetically manipulated micro-organisms and any biologically active products.

In line with the OECD report, the large scale guidelines express the view that the vast majority of large scale applications will use organisms of intrinsically low risk which warrant only minimal containment. This category of work is known as Good Large-Scale

Practice (GLSP) and the guidelines give criteria for categorising GLSP organisms consistent with those in the OECD report. The guidelines recognise, however, that it may in some cases be necessary to use organisms of higher risk, and they also provide risk assessment principles for such organisms together with examples of corresponding physical containment.

The European Commission draft *Proposal for a Council Directive on the Contained Use of Genetically Modified Micro-organisms* (European Commission 1988b) aims to harmonise the regulations existing in the different member states in relation to the contained use of genetically modified micro-organisms for laboratory and large scale activities. It provides for a system of notification, makes containment requirements according to the type of micro-organism used, and makes demands relating to accidents and waste management. Where an installation is to be used for the first time, a notification is required to a national Competent Authority. The draft directive establishes the classification of genetically modified micro-organisms into two groups according to the OECD criteria for GILSP (OECD 1986). Records have to be kept for low risk (GILSP) work and prior notificaion is required for work involving higher risk (non-GILSP) organisms.

Planned Release
For environmental and agricultural applications involving the release of genetically manipulated organisms the OECD report (1986) recognises the establishment of internationally agreed safety criteria would be premature. Accordingly, a provisional approach is recommended incorporating independent case-by-case review of the potential risks of such proposals. The report recommends the factors that the group felt should be taken into account during risk assessment, advocating that existing data on the introduction of alien organisms should guide our attitude to genetically altered ones. The report recommends a cautious approach to the release of rDNA organisms into the environment, recommending an incremental transition from the laboratory to a contained environment (for example a greenhouse) to small scale field tests.

In the UK the ACGM recognised that the issues raised by the planned release of genetically manipulated organisms into the environment should be considered as a priority. A working group was set up in 1984 and this led to the publication, in April 1986, of ACGM guidelines for the planned release of genetically manipulated organisms for agricultural and environmental purposes (ACGM/HSE/Note 3). The working group became the ACGM's specialist sub-committee – the Planned Release Sub-Committee [renamed the Intentional Introduction Sub-Committee]. The structure of the Planned Release Sub-Committee follows the tripartite structure of the ACGM, being composed of employers, employees and scientific specialists. It comprises not only the relevant scientific experts including ecologists, but also representatives from industry, trades unions, local authorities, the Public Health Laboratory Service, NERC, Forestry Commission, MAFF, DoE, DoH, HSE Inspectors and the NCC. (Table 23.1 lists the members of this Sub-Committee in 1988.) Thus proposals are not subjected merely to scientific or peer review.

The ACGM considers it to be essential that a Local Genetic Manipulation Safety Committee undertake initial risk assessment of a proposed planned release, which is then submitted as part of the proposal to its Planned Release Sub-Committee. Table 23.2 lists some of the factors which should be addressed in the initial risk assessment. These are an abridged version of what is required in the guidelines, and they indicate the considerable amount of information that the sub-committee considers for each planned release proposal.

The Planned Release Sub-Committee considers each planned release proposal in detail. The notifier is expected to be present at certain stages of the deliberations of the sub-committee in order to present his or her case and to answer questions. The sub-committee then has to arrive at a unanimous decision. The ACGM considers it essential that there be local consultations with groups like the local environmental health department and that there should be press releases and local announcements.

Learning by experience is an important element in the oversight of this area. Given the uncertainty that exists whether there are real risks associated with the release of genetically modified organisms into the environment it is important to be able to accommodate new knowledge, and adjust procedures accordingly. The sub-committee expects the proposers to report back on their work. Such feedback is considered to be vital to the further development of the guidelines and methods of risk assessment.

The planned release of genetically manipulated organisms into the environment in the UK is not just the remit of the HSE. In July 1989 the UK Royal Commission on Environmental Pollution published its report on *The Release of Genetically Engineered Organisms into the Environment*. This report is a consultative document to be used by government departments when they are considering how they should regulate planned releases. The HSE, the DoE and the DTI are each funding some risk assessment research in the area of planned release. The ACGM provides the forum for other government agencies to contribute to its deliberations on proposals for

Table 23.1:
Members of the Advisory Committee on Genetic Manipulation (ACGM) Planned Release Sub-Committee in 1988*

Title	Organisation
Chairman	
Professor J.E. Beringer	Microbiology Department and Unit of Molecular Genetics Bristol University
Members	
Dr H.D. Burges	Glasshouse Crops Research Institute
Dr M. Crawley	Department of Biology, Imperial College, London
Professor D. Ellwood	Formerly CAMR & Professor of Environmental Sciences, University of Warwick
Dr H.F. Evans	Forestry Commission
Mr C.R. Franks	Environmental Health Department, Borough of Brighton
Dr J.G. Jones	Freshwater Biological Association
Dr J. Kinderlerer	Department of Biochemistry, University of Sheffield
Professor D. Onions	Department of Veterinary Pathology University of Glasgow
Dr N.J. Poole	ICI
Dr M. Vincent	Nature Conservancy Council
Professor M. Williamson	Department of Biology, University of York
Assessors	
Dr J. Drozd	Biotechnology Unit, DTI
Mr T.E. Tooby	MAFF
Ms M. Pratt	MAFF
Dr C. Wray	MAFF
Mr P. Lister	DHSS
Dr H. Murrell	DHSS
Mr J.F.A. Thomas	DoE
Mr M.S. Chapman	HSE
Mr M. Devine	HSE
Secretariat	
Mr B.P. Ager	HSE
Ms J. Soave	HSE
Mr M. Bailey	HSE

Notes:
*	membership is not static
CAMR	Centre for Applied Microbiology Research, Porton Down
ICI	Imperial Chemical Industries
DTI	Department of Trade & Industry
MAFF	Ministry of Agriculture, Fisheries & Food
DHSS	Department of Health & Social Security; now the Department of Health (DOH)
DoE	Department of the Environment
HSE	Health & Safety Executive

Table 23.2:
Advisory Committee on Genetic Manipulation (ACGM)
risk assessment factors for the notification of a planned release
of genetically manipulated organisms

a.	The objectives of the project.
b.	The nature of the cell or organism to be released.
c.	The procedure used to introduce the genetic modification.
d.	The nature of any altered nucleic acid and its source, its intended function and the extent to which it has been characterised.
e.	Verification of the genetic structure of the novel organism.
f.	The genetic stability of the novel organism.
g.	The ability of the organism to give rise to long-term survival forms and the effect the altered nucleic acid may have on this ability.
h.	In the case of a pest control agent, details of the target biota.
i.	The size and nature of the site of release.
j.	The physical and biological proximity to humans and other significant biota.
k.	Details of the ecosystem into which the organism is to be released.
l.	The method and amount of release, rate, frequency and duration of application.
m.	Monitoring capabilities and intentions.
n.	The on-site worker safety procedures and facilities.
o.	The contingency plans in the event of unanticipated effects of the novel organism.
p.	Survival and persistence of the novel organism.
q.	An assessment of the environmental consequences of the release.

planned release, and to consider other legislative requirements that may apply and for which they are responsible. Such appropriate legislation could include pesticide regulations, the DoE's Wildlife and Countryside Act and the Plant Health Act.

The European Commission's draft *Proposal for a Council Directive on the Deliberate Release to the Environment of Genetically Modified Organisms* (European Commission 1988c) provides for a case-by-case notification and endorsement procedure. A distinction is made between the procedures for research and development releases, and releases involving finished products. In the latter case the endorsement procedure involves consultation with the commission and with other member states. The product section of the draft directive does not apply to organisms already covered by EC product legislation.

The draft directive proposes that before carrying out a release the responsible person has to notify the Competent Authority, giving details on the organisms proposed for the release, the conditions and the environment in which the release is to take place, and an assessment of the possible hazards to human health and the environment which may arise. The major points about this proposed piece of EC legislation are that the national Competent Authorities in the member states must conduct a case-by-case review of each planned release proposal using similar principles to those of the ACGM, listed in Table 23.2. Those EC Member States which do not already have national Competent Authorities will need to establish such bodies.

Conclusions

The regulatory systems for genetic manipulation implemented in many countries during the 1970s were broadly similar to each other. At that time genetic manipulation work was mainly laboratory based. As the focus began to shift to applications on a large scale and to planned releases into the environment, there was a need to re-examine regulatory requirements.

The OECD's 1986 report was timely because it established a scientific framework for the safe use of rDNA organisms in industry and in the environment at a stage when many national regulatory authorities were considering how to approach risk assessment in these areas. This internationally agreed framework should facilitate the development of safe biotechnology along common lines in many countries. Our ability to assess any risks should improve with increased knowledge and experience. The OECD has launched a follow-up study to its 1986 report which should assist further international harmonisation in regulating the use of genetically manipulated organisms. Once agreed and implemented, the EC draft directives will also harmonise the approach towards the regulation of this new technology in all the member states of the EC.

International understanding of the safety issues in this area not only provides the basis for a consensus on the protection of public health and the environment, but it also leads to the promotion of technological and economic development and the reduction of national barriers to trade.

24 Genetic Engineering in Europe

BENEDIKT HAERLIN

Disinterested Information

I should like to begin with a quote from the minutes of an OECD meeting held in Toronto in April 1987 on *National Policies and Priorities in Biotechnology*. The speaker was Dr Karl Heusler, Director of Research at Ciba-Geigy in Basle:

> Today it is possible to set into motion global processes which can neither be reversed nor reliably restrained, and which may end in the annihiliation of our planet. It is the ruling paradigm of scientific and technical 'progress' which has brought us to this level of potential destruction – a paradigm which dictates that whatever can be done *must* be done, and that it is only in the application of new knowledge that the consequences become evident. This notion of unrestrained 'progress' based on trial-and-error is clearly no longer adequate to the dimension of the problems we face today. We must develop a new ethic for dealing with our knowledge and cannot entrust this task only to scientists, politicians, and so-called experts, nor can we leave it to the mechanisms of the free market and international competition (OECD 1987).

Dr Heusler was quoting from the *Founding Appeal* of the Gen-Ethic Network, an organisation which I was pleased to participate in setting up in 1986. In quoting from this document he was trying to illustrate the formidable opposition from the public which his company is facing today in the field of what he calls 'biotechnology', but which I prefer to call 'genetic engineering'.

Dr Heusler named four principles necessary to achieve public acceptance in this field. First, the informant must be credible. Second, over-information leads to passive refusal. In other words, do not answer questions which have not been asked. Third, there are always two sides of the coin: there are benefits and there are risks. Fourth, information should not pre-empt decisions.

On the first principle for achieving public acceptance, Dr Heusler states that:

> In order to be credible the informer should not profit in any way nor suffer unduly from a decision in the areas in question. He should be well informed, but also be aware of the concerns of the public. Such persons are difficult to find. (OECD 1987)

Why are such persons so difficult to find? I have three answers to this question. First, we are dealing with a new and complex technology which is accumulating knowledge rapidly. Second, genetic engineering is a field with very high commercial expectations. Millions of dollars have been invested into research and development in genetic engineering in the past 15 years or so. The financial return on these investments is still awaited. Scientists who have knowledge in this field are sought throughout the world, and for every one such scientist there are at least two well-paid jobs. There are very few scientists in this field who are able or willing to provide disinterested advice to policy makers. Third, critics of genetic engineering not only lack financial resources, but also sometimes engender furious reactions and sanctions from the scientific community as well as from companies. This is because the basic assumption that genetic engineering does not involve qualitatively new and serious risks is only a few years old and is based on a rather fragile consensus within the scientific community itself. In the early 1970s the prevailing consensus was that genetic engineering is dangerous, and the onus of proof lay with the experimenters to demonstrate that their experiments were safe. It took an enormous effort on the part of the scientific community and investors to reverse this consensus (see Wheale and McNally 1988a, Chapter 3). Whoever challenges this reversal of the burden of proof not only faces the usual sanctions towards minority views within the scientific community, but he or she will also be confronted with sometimes hysterical reactions rather than with scientific reasoning. The hysterical response of the scientific community to critics of genetic engineering can be seen as an expression of the no-longer outspoken anxieties about its safety which many scientists have.

In the public debate about genetic engineering currently taking place it is difficult to know whom to believe, because there is very little independent knowledge available. Scientists who serve on expert commissions and who advise safety committees, for example, are usually directly involved in research in this field themselves, and their own research has to be approved and funded

We can expect there to be even fewer disinterested scientists in the future, as governments in industrialised countries, especially in the European Community (EC), are devising their science and technology policies to increase international competitiveness rather than safeguarding the traditional equilibrium between public and private control of technology. As a result of this change in emphasis, public funding has become virtually conditional upon the acquisition of industrial participation and direct co-operation. It is industry which is setting the scientific agenda in this young and technical field of genetic engineering, which lacks the tradition of responsibility for its innovations.

The EC biotechnology research programmes called the European Collaborative Linkage of Agriculture and Industry through Research (ECLAIR) and the Food-Linked Agro-Industrial Research (FLAIR) are perfect examples of this dangerous attitude. The objective of these programmes is to convert agriculture into a branch of industry. The ECLAIR programme has a budget of 80 million ECU (European Currency Units) to develop new plants that are custom-tailored to the needs of industry. ECLAIR promises farmers that they might be able to continue to over-produce by selling their produce to industry as well as to consumers. The FLAIR programme concentrates on food quality, competitiveness and the safety of foodstuffs, and on their nutritional and toxicological properties. Under FLAIR, a total of 25 million ECU is available for transnational projects with significant industrial participation. In my opinion the most striking task for FLAIR is to measure food quality objectively. The idea is to take certain compounds of food which can be measured, and to treat the measurements of these compounds as a measure of food quality.

Scientism and Genetic Engineering

The time taken nowadays to convert basic research into technological innovation is getting shorter as a result of technical as well as political reasons. Consequently, the time available to the public to assess technological innovations – culturally, politically and ethically – is also getting shorter. Yet the consequences, for example, of releasing genetically engineered organisms into the environment, or the reduction of the global gene pool due to the patenting of living material, have to be considered on a time scale which is potentially infinite.

Only today we realise the impact on the environment of using agrichemicals. Billions of dollars have to be spent to clean up pollution and prevent further deterioration of our environment. Only today do we realise the consequences of the exponential increase in energy consumption over the past 100 years or so. Only tomorrow, it seems, will we fully realise the terrible burden we have imposed upon ourselves and future generations through the exploitation of nuclear energy. Greater caution and an attempt to replace short-term and single-minded concepts of technical progress with more cautious, long-term and responsible approaches could be seen as rational responses to these experiences. The search for adequate rules of prospective evaluation, as well as appropriate ways of making decisions on research and technology, could be another rational response to the ecological disasters and challenges we face today. The further development and dissemination of research in the life sciences will undoubtedly play a crucial role in this context. Of course risk assessment in the narrow sense of predicting possible inconveniences associated with a given technique is not a very attractive field of research, but the development of predictive ecology to evaluate the risks of new technologies to avoid environmental problems could be attractive to young scientists.

Genetic engineering is a very powerful technology. The more powerful a technology is the better it has to be understood, and the more cautiously it has to be used. In order to accomplish this the goals, as well as the options, have to be clearly defined beforehand.

The list of new and beneficial products which genetic engineering promises is long and impressive. It includes herbicide resistant and pesticide resistant plants; biological pesticides; pest resistant and disease resistant plants and animals; new plants and animals with improved efficiency; new approaches to plant nutrition through the use of nitrogen-fixing microbes; task-tailored microbes for agriculture, environmental control and mineral leaching, and new vaccines.

The principles to which genetic engineering is dedicated are not new at all. They are the faster, the bigger, the better. The ends have not changed, only the means. Even the companies have not changed: the markets and strategies for genetic engineering are being determined by existing multinational petrochemical and agrichemical companies. Those who told us the 'Green Revolution' would eradicate pests and banish hunger from the world today offer new technologies for old promises with the same ruthless optimism which is derived from prospective profits rather than from social benefits. Why should we believe them? It is true that some of the promises are new, such as cleaning up the waste produced in the last 'Green Revolution', or coping with the problems stemming from other past technological 'revolutions'. However, just as political revolutions always tend, depressingly, to reproduce what they attempt to replace, so technological 'revolutions' tend simply to shift the shortcomings of older technologies to another level without solving the underlying structural problems.

The effect of shooting flies with cannons is well known. The effect of improving the storage properties of ice cream through the use of genetically engineered 'ice-plus' bacteria has yet to be discovered!

Genetic engineering should not be used as a form of 'scientism' to maintain and enforce industrial principles that create more problems than they solve. As long as genetic engineering is used to adapt nature to the short-term requirements of industrial production instead of industry adapting to the requirements of the equilibrium of nature, the only effect will be a shift, and most likely an increase, in the structural problems of human production and reproduction. Let me give you an example of what I mean by this: to abstain from producing more and more toxic waste is probably a more intelligent solution to the problem of environmental pollution than to create bugs that can digest it.

Europolitics

I am a Member of the European Parliament (MEP) and I would like to tell you something about the attitudes prevalent in the EC. Representatives of almost all political parties are anxious these days to express their deep belief in technological progress, and in particular in the improvement of the world through genetic engineering. The less politicians know about a technology, the easier they are convinced that our future depends upon it. Even when there are reservations about the safety of that technology, the best way to convince an average MEP is to tell him or her that the USA is ahead, Japan is on our heels, and the Pacific Basin is not sleeping.

Does the EC face a gene technology gap? In fact the EC leads in many fields of genetic engineering and is abreast of the developments in most others. However, this question leads to the same kind of vicious circle that disarmament has been stuck in for so long. It results in a ridiculous Euro-patriotic attitude amongst the political and administrative mainstream, namely, that the EC must be ahead – no matter where the journey will take us.

In this vicious political circle, technology no longer appears as the more or less useful neutral tool with which to achieve certain ends, but becomes the end itself. In political decision-making, the search for useful applications appears to be ancillary to, rather than the purpose of, new technologies.

The EC is playing an important role in developing pro-active regulations for genetic engineering [see Chapters 22 and 23]. The European Commission is trying to establish these regulations before national laws are implemented. The first Council Directive for the field of genetic engineering was adopted by the Council of Ministers in 1986. This directive aims to speed up the approval process for drugs derived from high technology especially in the field of biotechnology.

This directive assists the competitiveness of the EC biotechnology industry by facilitating its access to a bigger market. Ironically, the first companies to make use of this directive were two US-based multinationals. Their product, bovine somatotropin (BST) or bovine growth hormone (BGH), promises to increase milk yield in treated cows by up to 25 per cent [see Chapter 7], but increased milk production is not the European Economic Community's (EEC's) most pressing need! On the contrary, one of its priorities is to reduce the size of the 'milk lakes' and other food surpluses, and thereby reduce the absurd level of the Common Agricultural Policy (CAP) budget. The most likely effect of marketing BST will be to further the concentration and intensification in agriculture. According to US studies, two-thirds of small dairy farmers will go out of business within the first ten years of its application. The health and fertility of dairy cows will suffer from the application of BST, and the result will be additional applications of antibiotics and other veterinary pharmaceuticals. The quality of milk and the consumer's confidence in milk products will be put at risk [see Part II].

The European Parliament, together with the majority of Ministers of Agriculture from milk-producing EC Member States, have expressed concern over BST. The broad opposition to this new generation of hormones has not been appeased, despite political lobbying by the chemical and pharmaceutical industry on both sides of the Atlantic, organised in the EC by the European Federation for Animal Health. Yet, the European Commission is backing BST, the main reason being that the failure of BST to reach the market would be a discouraging signal to the genetic engineering industry. Half a billion dollars has already been invested in the development and promotion of BST; it is one of the first products of genetic engineering in the agricultural field and it has a potential multi-million dollar market.

The case of BST illustrates three typical EC approaches to technology. First, it illustrates the narrow-minded concept of competitiveness. Companies appeal to the patriotic short-sightedness of the EC, expressed as 'If we don't do it, others will do it before us'. The EC response is deregulation. However, in genetic engineering, as in many other technological fields, deregulation serves multinational companies best. Secondly, the BST case is a fine example of technology in search of a market. Thirdly, it demonstrates the remoteness of the European Commission in Brussels from public opinion. If the European Commission had to stand for elections it would surely react more reasonably and sensibly. The only real pressure on the commission, apart from governmental, is from the well-organised industrial pressure groups. Industrial pressure groups are highly-funded (US$20 million are said to have been spent on the BST public acceptance campaign) and well integrated into the decision-making and daily administration of the European Commission. The administrative partner of the genetic engineering industry in the European Commission is the Concertation Unit for Biotechnology in Europe (CUBE), designed to co-ordinate the activities of the various EC agencies involved in genetic engineering.

In the spring of 1988 the European Commission published three proposals for Council directives on the deliberate release of genetically manipulated organisms into the environment, the contained use of such organisms, and the protection of workers from the risks related to the exposure to biological agents at work (European Commission 1988a; 1988b; 1988c).

Commenting on the proposed directive on the deliberate release of genetically modified organisms, the European Commission states that this proposal has been discussed with the European Biotechnology Coordinating Group (EBCG), composed of representatives from different European industries' organisations. The employee's side, however, has not been consulted. Industry's view is that Community-wide harmonisation of procedures is badly needed, but it expressed its reservations to an endorsement by regulatory bodies prior to the releases. This latter point is alarming because, as the proposed directive currently stands, the potential of the Competent Authorities [see Chapter 23] of the Member States effectively to control the deliberate release of genetically modified organisms is minimal. Once notification of intent to undertake a deliberate release has been given to the Competent Authority of a Member State, a decision which will cover the whole EC has to be made within 90 days, provided the Competent Authority of any other Member State does not object. In cases where there are disagreements between Member States over such decisions, the commission wants to have the right of final endorsement.

There are no provisions in the proposed directive on deliberate release for public information and participation in the decision-making process. The commission states that in a field largely unknown like this the decisions on the safety of a release and its conditions must be the result of a dialogue between notifier and Competent Authority. However, until recently some Member States have not had a Competent Authority, and such Member States, being eager to catch up with this new technology, may well become the favourite places for the deliberate release of genetically engineered organisms by multinational companies. In effect the proposed directive, which was supposed to set maximum rather than minimum levels of safety regulations, will instead ensure the legal deregulation of safety requirements in this field.

In addition to the above proposed directives, the European Commission has issued a proposed directive designed to harmonise the patent law in genetic engineering (European Commission 1988d). The European Commission's Industry Research and Development Advisory Committee (IRDAC), comprising representatives of the genetic engineering industry, has pointed out that the harmonisation of legislation for the patenting of biotechnological inventions is indispensible to ensure favourable conditions for a common market in genetically altered organisms and products. The key statement of the proposed directive is: 'A subject matter of an invention shall not be considered unpatentable for the reason only that it is composed of living matter' (European Commission 1988d, Article 2).

Under this proposed directive, animals and plants will become patentable if they are altered by genetic engineering. Thus this proposed directive will be a strong incentive to use genetic engineering rather than traditional breeding methods, which do not render their subjects patentable. The patentability of living material places our global genetic resources and genetic diversity within the realm of corporate ownership [see also Part I].

Predictive Medicine

The European Commission's latest proposal for genetic engineering is to coordinate research on mapping and sequencing the human genome (European Commission 1988e). The commission promises that this research programme will result in a 'Europe of health' with decreased social and healthcare costs. The subtitle of the research proposal – *Predictive Medicine* – tells us a lot about its underlying concept of genetic determinism, and its intention to discriminate against those apparently susceptible to genetic disorders [see also Chapter 21]. The reasons for the research proposal are given as follows:

Fifty years ago the principal cause of morbidity and mortality was infectious disease but with the discovery of antibiotics, and improvements in hygiene and pest control, it is now a minor one in industrialized countries. Apart from the consequences of accident or war, much disease today has a genetic component which may be of greater or lesser importance ...

However, when it comes to the common diseases such as coronary artery disease, diabetes, cancer, autoimmune diseases, the major psychoses and other important diseases of Western society, the position is far less clear. These conditions have a strong environmental component and although genetic factors are undoubtedly involved, they do not follow any clear-cut pattern of inheritance. Put another way, the disease results from the exposure of genetically susceptible individuals or populations to environmental causes ... As it is most unlikely that we will be able to remove completely the environmental risk factors, it is important that we learn as much as possible about the genetically determined predisposing factors and hence identify high-risk individuals. In summary, Predictive Medicine seeks to protect individuals from the kinds of illnesses to which they are genetically most vulnerable and, where appropriate, to prevent the transmission of the genetic susceptibilities to the next generation (European Commission 1988e, p. 3).

Conclusions

Genetic engineering will be a key technology of the 21st century. It will be the first new technology to emerge under the conditions of a common European market after 1992. There is little hope that the existing EC structure and its policies will guide and control the development of this technology so that it works in harmony with nature to serve the needs of human society, rather than dominate and impose its reductionist technological principles on human societies and the environment. We can hardly imagine the extent of the changes necessary to render the EC bureaucracy and administration democratic. We need international regulations and a strong public counterpart to the power of the multinational genetic engineering industry.

Discussion V

Anon: I think that the sequencing of the human genome is one of the most exciting projects at the present time and has immense value for pure research. I should like to ask Edward Lee Rogers if he thinks it should be stopped?

Edward Lee Rogers, Foundation on Economic Trends: I do not think that anyone would argue that there should not be selective research to isolate the genetic defects responsible for specific inherited disorders. But that is a far cry from the objectives of the US Human Genome Project, which is to sequence and map the entire human genome. In terms of research priorities this is not an appropriate thing to do because we have other more pressing priorities and research needs. The Foundation on Economic Trends opposes the Human Genome Project primarily on ethical grounds because there is no mechanism in place to protect the right of privacy, and that would be true even in the case of selective research into the human genome.

Dr Wolfgang Goldhorn, State Veterinary Service, FDR: Does Edward Lee Rogers think that the amount of voluntary genetic screening will increase as job applicants find that a good genetic screening profile increases their prospects of gaining employment?

Edward Lee Rogers, Foundation on Economic Trends: I think that there is going to be a great deal of voluntary genetic screening. There will be people who, feeling confident of their genetic profile, will want to demonstrate it to prospective employers and insurance companies and others.

Anon: Why do scientists who know what is happening in the field of genetic engineering and are sympathetic to animals not speak out more strongly against this technology? I believe that the animal suffering caused through genetic engineering will lead to human suffering in the future. Perhaps Dr Tait would like to comment on this view.

Dr Joyce Tait, Open University: That which worries the general public also worries people who work in industry. I think if you look at the past history of risk regulation, in areas like pesticides, for example, you will always find a few industrial scientists actively involved with the products being developed who spotted potential problems before the public could have. It is very important that we actively encourage the free flow of that kind of information. We should encourage people who are involved with new technology to express their concerns and worries, perhaps anonymously so they will not suffer sanctions at work. People who work in regulatory agencies should be similarly encouraged. We need anonymous surveys of people from industry, regulatory agencies and the public to provide us with information which will enable us to control new technologies more effectively. I agree that control of genetic engineering and biotechnology is going to be very important, and we should increase our vigilance in this area as time goes by and the number of products available increases.

Baroness Edmi di Pauli: Would the panel comment on the suggestion of the Institute of Social Invention that an ethical oath, equivalent to the medical profession's Hippocratic Oath, should be introduced for scientists and technicians? May I quote the last sentence of the proposed oath: 'I will not permit considerations of nationality, politics, prejudice or material advancement to intervene between my work and this duty to present and future generations.'

Benedikt Haerlin, MEP: I have never heard the suggestion of an ethical oath for scientists before but I think it is an excellent idea. It would help to shield scientists from the enormous pressure which is on them these days to undertake research which they feel is unethical. I should be interested to learn the full wording of this proposed oath and I would be happy to promote this idea wherever I can.

Dr Caroline Murphy, RSPCA: I also find this an interesting proposal and would certainly support it. I think it would be marvellous if scientists were placed under a series of constraints, as are medics.

Dr Peter Wheale, Bio-Information (International) Limited: I think the proposed ethical oath for scientists and technicians is interesting. However, even with the Hippocratic Oath, there is a problem in medicine which is worth bearing in mind. This problem is that the drugs that doctors administer to patients are researched, developed and manufactured by scientists and technicians who are not subject to the Hippocratic Oath. Similarly, because the practice of science and technology does not constitute a profession as such, it would be very difficult to impose sanctions on those scientists and technicians who violated the proposed ethical oath.

Scientists and technicians in the public sector, however, could be asked to sign such an ethical oath as a condition of their employment contract, and sanctions could then be imposed on those who violated it. Public pressure could be brought to bear on companies to persuade them that it would make economic sense to adopt a policy of adhering to such an ethical code. Companies with an approved code of ethics, including an ethical oath, could perhaps be permitted to join a recognised organisation of ethically responsible companies, from which they could be excluded if they were found to have violated this ethical code.

Christine Cremer: Would any of the contributors care to comment on the suggestion that the human immunodeficiency virus (HIV), which is believed to be responsible for acquired immune deficiency syndrome (AIDS), was created through genetic engineering.

Mark Cantley, CUBE: The overwhelming consensus amongst scientists is that it is inconceivable that anybody could have been clever enough to make such a sophisticated virus. Furthermore, there are blood serum samples from the 1960s stored in Norwegian laboratories which have traces of HIV in them. This suggests that HIV arose prior to the advent of genetic engineering.

Benedikt Haerlin, MEP: I believe that it might be possible to genetically engineer such a virus but I am not convinced by any of the arguments which have been brought up in favour of this theory.

Ruth McNally, Bio-Information (International) Limited: The origin of HIV or, more accurately, the HIVs and related viruses in other primates, is the subject of debate, and whether HIV is the causal agent of AIDS has been disputed. The National Anti-Vivisection Society (NAVS) has published a report called *Biohazard* [NAVS 1987]. In this report they reference sources which support the thesis that HIV could be the result of vivisection experiments carried out in the USA and Europe. The report concludes that whether or not HIV was created in the laboratory is less important than the fact that it is possible to create such a virus through vivisection work conducted without due regard to the biohazards involved.

I should also like to mention a couple of points on the subject of the origin of HIV which were raised in a letter to *Nature* from John Seale of Harley Street [Seale 1988]. He notes that the nucleotide sequence of HIV-2 resembles the nucleotide sequence of SIV_{MAC} [simian immunodeficiency virus of macaque monkeys] more than it resembles HIV-1. That is, a human immunodeficiency virus and a monkey immunodeficiency virus resemble each other more than do two human immunodeficiency viruses. This evidence is consistent with the view that these viruses have jumped species. Furthermore, SIV_{MAC} has never been found in wild populations of macaques; it has only been observed in primate research colonies. Seale's conclusion is:

The fact that it is possible, indeed quite simple, for HIV-1 and HIV-2 to have been created by artifical selection, and the AIDS epidemic started deliberately, does not necessarily mean that this actually happened. But it is a possibility that deserves serious critical analysis by the scientific community – in public [Seale 1988].

Anon: One of the most enduring images that I shall take from this conference is the total disagreement at almost every level between the proponents of genetic engineering technology and those who feel concerned about its implications. In the light of the almost complete irreconcilability of these two groups, I would like to ask Mark Cantley, as a regulator working for the European Commission, are there any grounds for compromise or will one view or the other prevail?

Mark Cantley, CUBE: I do not share the sense of hopelessness that seems to be implicit in that question. I think that reconciling apparently irreconcilable differences is our daily business in Brussels – differences within and between the European Commission, the European Parliament, the Council of Ministers and, occasionally, if things remain unresolved, the European Court. The function of politicians is to find acceptable compromises, and people faced with making joint decisions do not have to agree on their philosophy of life. They do not have to agree at the level of 'lovely ideas' to find a pragmatic compromise. There has been quite a lot of exaggeration about the threats and the potential of genetic engineering. I suggest that we let the legislation and regulation evolve *pari passu* [apace] with the development of genetic engineering.

The Reverend Dr Andrew Linzey, Essex University: The problem with Mark Cantley's suggestion of allowing the controls over the practice of genetic engineering to develop with the technology is where this will lead us. The danger is that if something has been done in the past, or is allowed to go on happening for a few years to see what will happen, it tends to become a tradition, a ritual, it becomes the way in which things have to happen, and becomes inexorable.

Mark Cantley, CUBE: As Bernard Shaw said: 'The reasonable man adapts himself to the world: the unreasonable one persists in trying to adapt the world to himself. Therefore all progress depends on the unreasonable man.' That, to me, is the definition of science. The scientist has the humility to admit that he does not know and the arrogance to try and find out. The clerics did not want to look through Galileo's telescope! The existence of technology can alter people's freedom but that is a risk that has to be taken. People have the option of putting the knowledge aside, but on the whole, our progress over the last two or three centuries has been based on having sufficient confidence to assimilate novelty and not being too frightened of knowledge *per se*. My point is that we need good controls on its applications.

Joyce D'Silva, the Athene Trust: Would Mark Cantley tell us about the regulatory consultation on genetic engineering which is now taking place in the EC?

Mark Cantley, CUBE: I estimate that there are about 300 existing Council directives relevant to biotechnology, mainly in the areas of foods and drugs. There are also four proposed directives which relate to genetically modified organisms – on contained use, deliberate release, worker health and safety, and patenting [European Commission 1988a, 1988b, 1988c, 1988d]. These proposed directives will be debated for many months in parliamentary committees and in Council working groups. If groups or individuals have strong views on these proposed directives or interesting points they would like to have considered in the discussions on them, I suggest they write to their MEP who will be happy to transmit their comments to the appropriate desk within the European Commission.

Benedikt Haerlin, MEP: Politics is the art of compromise, but who sets up the agenda? For example, there was no democratic debate on whether or not we want to release genetically engineered organisms into the environment. You should all be aware of the fact that there is no democratic decision-making within the EC. If you want to have an impact on decisions relating to the proposed directives on genetic engineering, you should lobby your national government, because it is their representative in the Council of Ministers who makes the final decision, not the European Parliament.

References V

ACDP (1984) *Categorisation of Pathogens According to Hazard and Categories of Containment* (London: HMSO).

ACGM/HSE/Note 1 *Guidance on Construction of Recombinants Containing Potentially Oncogenic Nucleic Acid Sequences* (London: HSE).

ACGM/HSE/Note 2 *Disabled Host/Vector Systems* (later revised and annexed with Note 7) (London: HSE).

ACGM/HSE/Note 3 *Guidelines for the Planned Release of Genetically Manipulated Organisms for Agricultural and Environmental Purposes* (London: HSE).

ACGM/HSE/Note 4 *Guidelines for the Health Surveillance of Those Involved in Genetic Manipulations at Laboratory and Large-Scale* (London: HSE).

ACGM/HSE/Note 5 *Guidance on the Use of Eukaryotic Viral Vectors in Genetic Manipulation* (London: HSE).

ACGM/HSE/Note 6 *Guidelines for the Large-Scale Use of Genetically Manipulated Organisms* (London: HSE).

ACGM/HSE/Note 7 *Guidelines for the Categorisation of Genetic Manipulation Experiments* (London: HSE).

ACGM/HSE/Note 8 *Laboratory Containment Facilities of Genetic Manipulation* (London: HSE).

ACGM/HSE/Note 9 *Guidelines on Work With Transgenic Animals* (London HSE).

Advisory Council for Applied Research and Development (1980) *Biotechnology* (London: Advisory Board for Research Councils, Royal Society Joint Report, HMSO).

Ajzen, I. and Fishbein, M. (1980) *Understanding Attitudes and Predicting Social Behaviour* (Englewood Cliffs, New Jersey: Prentice-Hall).

Bernal, J.D. (1969) *Science in History* Volume 1 (Harmondsworth: Penguin).

Bishop, D.H.L. *et al.* (1988) 'Field trials of genetically engineered baculovirus insecticides' in M. Sussman *et al.* (eds) *The Release of Genetically Engineered Microorganisms* (London: Academic Press).

Cantley, M. (1988) 'Biotech safety regulations and public attitudes in the EEC', *World Biotech Report* 1988, Proceedings of Biotech88 Conference London, May 1988. Online Publications, pp. 77–88.

Davis, B.D. (1987) 'Bacterial domestication: Underlying assumptions' *Science*, vol. 235, pp. 1329–1332.

DoE (1989) *Proposals for Additional Legislations on the Intentional Release of Genetically Manipulated Organisms* (London: DoE).

Eden, C. *et al.* (1983) *Messing About in Problems* (Oxford: Pergamon Press).

Ellul, J. (1965) *The Technological Society* (London: Cape).

European Commission (1986) *A Community Framework for the Regulation of*

Biotechnology (Brussels: Communication from the Commission to the Council, CEC) (Brussels: COM (86) 573).

European Commission (1988a) *Proposal for a Council Directive on the Protection of Workers from the Risks Related to Exposure to Biological Agents at Work* (Brussels: COM (88) 165 final – SYN 129).

European Commission (1988b) *Proposal for a Council Directive on the Contained Use of Genetically Modified Micro-Organisms* (Brussels: COM (88) 160 final – SYN 131).

European Commission (1988c) *Proposal for a Council Directive on the Deliberate Release to the Environment of Genetically Modified Organisms* (Brussels: COM (88) 160 final – SYN 131).

European Commission (1988d) *Proposal for a Council Directive on the Legal Protection of Biotechnological Inventions* (Brussels: COM (88) 496 final – SYN 159).

European Commission (1988e) *Proposal for a Council Decision Adopting a Specific Research Programme in the Field of Health: Predictive Medicine: Human Genome Analysis* (1989–1991) (Brussels: COM (88) 424 final – SYN 146).

Gervers, J.H. (1987) 'The NIMBY Syndrome: Is it inevitable?', *Environment*, vol. 29, no. 8, pp. 18–20.

GMAG Note 14 *Revised Guidelines for the Categorisation of Recombinant DNA Experiments* (London: HMSO).

Habermas, J. (1971) *Towards a Rational Society* (London: Heinemann).

HSC (1987) *Review of the Health and Safety (Genetic Manipulation) Regulations 1978* (London: HSE).

Juma, C. (1989) *The Gene Hunters* (London: Zed Books).

Lindberg, L.N. and Scheingold, S.A. (1970) *Europe's Would-be Polity* (New Jersey: Prentice-Hall).

Mapping and Sequencing the Human Genome (1988) Report of the Board of Basic Biology Commission on Life Sciences, National Research Council (Washington DC: National Academy Press).

NAVS (1987) *Biohazard* (London: NAVS).

Nelkin, D. (ed.) (1985) *The Language of Risk* (Beverley Hill: Sage Publications).

OECD (1986) *Recombinant DNA Safety Considerations: Safety Considerations for Industrial, Agricultural and Environmental Applications of Organisms Derived by Recombinant DNA Techniques* (Paris: OECD).

OECD (1987) *'National Policies and Priorities in Biotechnology'*, Minutes of Meeting, Toronto, April, Broschure Seite 110.

OSTP (1986) 'Co-ordinated Framework for the Regulation of Biotechnology', *Federal Register*, vol. 51, No. 123, pp. 23,301–50, 26 June.

OTA (1987) *New Developments in Biotechnology – Background Paper: Public Perceptions of Biotechnology* (Washington DC: US Government Printing Office OTA-BP-BA, May).

Partridge, P.H. (1963) 'Some notes on the concept of power', *Political Studies*, vol. XI, pp. 107–25.

Poole, N.J. *et al.* (1988) 'The involvement of European industry in developing regulations', in *Planned Release of Genetically Engineered Organisms (Trends in Biotechnology/Trends in Ecology and Evolution special Publication)* (Hodgson, J. and Sugden, A.M. eds), (pp. S45–7), Elsevier Publications, Cambridge.

RCEP (1989) *The Release of Genetically Engineered Organisms to the Environment* (London: HMSO).

Research Surveys of Great Britain Ltd (1988) *Public Perceptions of Biotechnology: Interpretative Report* (RSGB Ref. 4780 March).

Roe v. Wade (1973) 13 93 S Ct 705.

Royal College of Obstetricians and Gynaecologists Working Party (1985) 'Foetal viability in clinical practice' (London: Royal College of Obstetricians and Gynaecologists).

Schmid, G. (1989) 'Report on the proposal for a Council directive on the deliberate release to the environment of genetically modified organisms (COM (88) 160 final – Doc. C 2–73/88 – SYN 131) *European Parliament Committee on the Environment, Public Health and Consumer Protection* (Brussels: DOC-EN\PR\65098.TO PE 128.472/fin, 28 April).

Seale, J. (1988) ' "Artificial" HIV?', *Nature*, vol. 335, p. 391.

Tait, J. (1983) 'Pest control decision making on Brassica crops', in T.H. Coaker (ed.), *Advances in Applied Biology*, vol. 8 (London: Academic Press) pp. 122–88.

Tait, J. (1988) 'Public perception of biotechnology hazards', *Chemical Technology and Biotechnology*, vol. 43, pp. 363–72. US National Academy of Science (1986) *Introduction of Recombinant DNA Engineered Organisms into the Environment – Key Issues* (Washington DC: National Academy Press).

Webster v. Reproductive Health Services (1989) 57 USLW 5023.

Wheale, P.R. (1986), 'Politics of science and technology', in C. Boyle *et al.*, *People, Science and Technology* (Hemel Hempstead: Wheatsheaf and New Jersey: Barnes and Noble Books).

Wheale, P.R. and McNally, R. (1988a) *Genetic Engineering: Catastrophe or Utopia?* (Hemel Hempstead: Wheatsheaf; New York: St Martin's Press).

Wheale, P.R. and McNally, R. (1988b) 'Technology assessment of a gene therapy', *Project Appraisal*, vol. 3, no. 4, December, pp. 199–204.

Wright, S. (1986) 'Molecular biology or molecular politics? The production of scientific consensus on the hazards of recombinant DNA technology', *Social Studies of Science*, vol. 16, pp. 593–620.

Conference Resolutions

1 By a majority vote, those attending the Athene Trust International Conference (1988) endorse the following resolution concerning the marketing of milk and dairy products derived from BST-treated cows:

> Whereas BST has not been given government approval for use in commercial dairy cattle, and whereas the social and economic, environmental and animal welfare consequences of the use of BST have not been resolved, and whereas the European Parliament in an Urgent Resolution on September 16th 1988 demanded a worldwide ban on the production, marketing and application of genetically engineered growth hormones, be it therefore resolved that the current practice of mixing milk and dairy products from cows test-treated with BST with the milk and dairy products of untreated cows be immediately prohibited.

2 By a majority vote, those attending the Athene Trust International Conference (1988) endorse the following resolution concerning the development of genetic engineering:

> Whereas this conference is deeply concerned about the development of genetic engineering and its ethical implications, and whereas genetic engineering will have profound effects on humanity, animals and the environment, this conference:

i   calls for a moratorium on the deliberate release of genetically engineered organisms into the environment;
ii   moves that genetically engineered living organisms and animals should not be patentable;
iii calls for a ban on the transgenic manipulation of animals;
iv   stresses the necessity and urgency of a global debate to address the ethical, environmental, economic and social implications of genetic engineering.

271

Glossary

cf. (confer) compare
Gr. Greek
L. Latin

Acetonemia: A morbid state, marked by the presence of acetone in the blood; symptomatic of diabetes.
Acidosis: Increased acidity of the blood; a symptom of acute diabetes mellitus.
Ahimsa: Buddhist doctrine of not hurting; compassion, especially for animals.
Allele/allelomorph: (Gr. *allelon*, one another; *morphe*, form.) Alternative forms of a gene for a given trait.
Amino acids: The building blocks of protein structure; twenty different amino acids are commonly found in living organisms.
Amniocentesis: A prenatal screening procedure, usually carried out at around seventeen weeks of pregnancy, in which a few millilitres of the amniotic fluid and floating cells surrounding the foetus are withdrawn through a needle inserted through the mother's abdomen and uterine wall.
Anthropocentric: (Gr. *anthropos*, a man; L. *centrum*, centre.) Taking mankind as the pivot of the universe.
Antibiotic: (Gr. *ante*, against or opposite; *bios*, life.) A substance capable of killing or preventing the growth of a micro-organism; can be produced by another micro-organism or synthetically.
Antibody: A protein produced in the body in response to the presence of a foreign chemical substance or organism (antigen), shaped to fit precisely to the antigen and in such a way as to annul its action or help to destroy it; part of the body's defence (immune) system.
Antigen: A substance which causes the immune system of the body to manufacture specific antibodies that will react with it.
Artificial insemination (AI): The application of sperm to an unfertilised egg by means other than ejaculation during sexual intercourse.
Assay: A technique that measures a biological response, for example, a plaque assay measures the number of infective viruses suspended in a measured volume of medium (see also PLAQUE).
Autosome: A chromosome other than a sex chromosome.
Bacteria: (Gr. *bacterium*, a little stick.) The simplest organisms that can reproduce unaided; a class of single-celled micro-organism found living freely in water, the soil and in the air, and as parasites within plants and

animals; *Escherichia coli* (*E. coli*) is the species most commonly used as a host cell in recombinant DNA work.

Baculovirus: (L. *baculum*, rod). Rod-shaped DNA virus which is believed only to infect the cells of invertebrate animals.

Base: Part of the building blocks of nucleic acids, the sequence of which encodes genetic information; cytosine (C), guanine (G), adenine (A) and thymine (T) are the bases in DNA; C, G, A and uracil (U) are the bases in RNA (see DNA BASE PAIR).

Biochemistry: The study of the chemistry of living things.

Biogenetic waste: Biological waste that contains genetically modified organisms; includes sewage, refuse and effluent from biotechnological processes.

Biological containment: The use of organisms in genetic engineering applications and research which are genetically engineered so as to minimise their ability to survive, persist or replicate; also applies to the use of genetically deficient cloning vectors, which are deficient in their ability to move to a new host strain; also known as 'genetic enfeeblement' and 'crippling'.

Biologicals/biologics: Term used to describe drugs which are based on substances found in living animals, for example, insulin.

Bioprocess: A process that uses complete living cells or their components, such as enzymes, to provide goods or services, for example, brewing.

Bioreactor: Vessel in which a bioprocess takes place, for example, a fermenter.

Biosphere: (Gr. *bios*, life; *sphaira*, a ball or a globe.) All life that is encompassed under the vault of the sky; the part of the earth that is inhabited by living organisms; the earth's surface and the top layer of the hydrosphere (water layer) have the greatest density of living organisms.

Biota: The fauna and flora of a region.

Biotechnology: After Bull *et al.* (1982), the application of scientific and engineering principles to the processing of materials by biological agents to provide goods and services.

Cartesian: Pertaining to the French philosopher Rene Descartes (1596–1650), or his philosophy; Cartesian philosophy expounds mind–body dualism, ethical dualism and explanatory dualism.

Cell: (L. *cella*, a little room, from *cello*, hide.) In 1655 Robert Hooke (1625–1702), curator of the Royal Society, used the term 'cell' to describe the small, closed cavities he found upon microscopic examination of the outer bark of an oak tree; although the structures he observed were actually cell walls, which are absent from animal cells, the term has persisted to describe the basic unit of structure of all living organisms excluding viruses; as defined by Max Schultze (1825–74) a cell is 'a lump of nucleated protoplasm'; a generalised description of a cell is a mass of jelly-like cytoplasm, contained within a semi-permeable membrane, and containing a spherical body called the nucleus; a single cell constitutes the entire organism of a single-celled creature such as a bacterium; a human being is composed of million of cells.

Cell fusion: The fusing together of two or more cells to produce a single hybrid cell (see also HYBRIDOMA TECHNOLOGY and MONOCLONAL ANTIBODIES).

Chimaera/chimera: (L. *chimaera*, a mythological monster with the head of a lion, the body of a goat, and the tail of a dragon, vomiting flames.) An organism, cell or molecule (for example, DNA) constructed from material from two different individuals, or species.

Chorionic villus sampling (CVS): A prenatal screening procedure whereby cells of the chorion – the membrane which surrounds the embryo – are withdrawn for genetic analysis; the cells of the chorion are derived from the fertilised egg and contain the same genetic information as those of the foetus; CVS can be performed at any stage of pregnancy from about eight weeks of gestation onwards.

Christmas disease: see HAEMOPHILIA.

Chromosomes: (Gr. *chroma*, colour; *soma*, body.) Darkly staining structures, composed of DNA and protein, which bear and transmit genetic information; they are found in the nucleus of eucaryotic cells and free in the cytoplasm in the cells of procaryotes; the number of chromosomes in each somatic cell is characteristic of the species; in human beings the normal chromosomal constitution is twenty-two pairs of autosomal chromosomes, and one pair of sex chromosomes.

Clone: A collection of genetically identical molecules, cells or organisms which has been derived (asexually) from a single common ancestor.

Cloning: Making identical copies of biological entities – molecules, cells or individuals; hybridoma technology is a cloning technique.

Codon: Three successive nucleotides (or bases) which specify a particular amino acid or a 'punctuation mark' in the genetic code.

Colony (of micro-organisms): A dense mass of micro-organisms produced asexually from an individual micro-organism.

Congenital disorder: A malfunction which is present at birth; the term describes all deformities and other conditions that are present at birth whether they are inherited or newly arisen as a result of adverse environmental factors or transgenic manipulation of the embryo.

Conjugation: (L. *cum*, together; *jugare*, to yoke.) Term used to describe mating between bacteria.

'Crippled' (virus): see BIOLOGICAL CONTAINMENT.

Cytogenetics: The area of study that links the structure and behaviour of chromosomes with inheritance.

Cytoplasm: (Gr. *kytos*, hollow; *plasma*, mould.) The living contents of a cell excluding the nucleus.

Diffusion (of an innovation): The spread of an innovation, with or without modification, through a population of potential users.

DNA: Deoxyribonucleic acid; the molecule which for all organisms except RNA viruses encodes information for the reproduction and functioning of cells, and for the replication of the DNA molecule itself; information encoded in DNA molecules is transmitted from generation to generation.

DNA base pair: A pair of DNA nucleotide bases; one of the pair is on one chain of the duplex DNA molecule, the other is on the complementary chain; they pair across the double helix in a very specific way: adenine (A) can only pair with thymine (T); cytosine (C) can only pair with guanine (G); the specific nature of base pairing enables accurate replication of the chromosomes and helps to maintain the constant composition of the genetic material.

DNA fingerprinting: A technique which uses DNA probes to generate personal genetic profiles which are as specific to individuals as conventional fingerprints.

DNA probe: A short piece of DNA that is used to detect the presence of a complementary piece of DNA in a sample of DNA under analysis; used in genetic screening, for example.

DNA sequence: The order of base pairs in the DNA molecule; genetic information can be encoded in the sequence of bases.

DNA technology: See MICROGENETIC ENGINEERING.

Dominant genetic disorder: (L. *dominans*, ruling.) A disorder which is expressed in the phenotype when the gene responsible is inherited from just one parent (cf. RECESSIVE GENETIC DISORDER).

Double helix: The name given to the structure of the DNA molecule, which is composed of two complementary strands which lie alongside and twine around each other, joined by cross-linkages between base pairs (see DNA BASE PAIR).

Down-stream process: A process in industrial biotechnology which occurs after the bioconversion stage; for example, product recovery, separation and purification.

Duplicative transposition: The process whereby transposable genetic elements move around genomes; a copy of a transposable genetic element located on a chromosome is duplicated and then deposited at a new location without loss of the original sequence.

EC Directive: EC Community law; binding on Member States as to ends but not means; can be issued by the European Commission or the Council of Ministers.

Ecology: Term coined in 1866 by Ernst Haeckel (1834–1919) to describe the branch of biology dealing with inter-relations between organisms and their environment.

Ecosystem: (Gr. *oikos*, home; L. *systema*, an assemblage of things adjusted into a regular whole.) A unit made up of all the living and non-living components of a particular area that interact and exchange materials with each other.

Elution: (L. *elutio*, wash.) Washing away impurity; cleanse.

Enablement requirement: A patentor's legal obligation to provide full technical details of his or her novel process or product which should allow a person with ordinary skill in the field to duplicate the invention.

Enzyme: (Gr. *en*, in; *zyme*, leaven.) A biological catalyst produced by living cells; a protein molecule which mediates and promotes a chemical process without itself being altered or destroyed; enzymes act with a given compound, the substrate, to produce a complex, which then forms the products of the reaction; enzymes are extremely efficient catalysts and very specific to particular reactions; the active principle of a ferment.

Epizootic: (Gr. *epi*, upon; *zoon*, animal.) Disease affecting a large number of animals simultaneously, corresponding to 'epidemic' in humans.

Escherichia coli (*E. coli*): A bacterial species that inhabits the intestinal tract of most vertebrates and on which much genetic work has been done; some strains are pathogenic to humans and other animals; many non-pathogenic strains are used experimentally as hosts for recombinant DNA.

Eucaryotes: Cells or organisms whose DNA is organised into chromosomes with a protein coat and sequestered in a well-defined cell nucleus; all living organisms except bacteria and blue-green algae are eucaryotic (cf. PROCARYOTES).

Eugenics: (Gr. *eu*, well; *genos*, birth.) After Galton (1883), the science which deals with all the influences that improve the inborn qualities of a race; also with those that develop them to the utmost advantage.

Evolution (biological): (L. *evolvere*, to unroll.) Changes in DNA that occur during the history of organisms; the development of new organisms from pre-existing organisms since the beginning of life.

Expression (of genes): See GENE EXPRESSION.

Factor VIII: One of approximately thirteen substances – or factors – involved in blood-clotting; haemophilia A (classical haemophilia), a blood-clotting disorder, is caused by a deficiency of Factor VIII and is treated by administration of exogenous Factor VIII (see also HAEMOPHILIA).

Factor IX: A blood-clotting factor; exogenous Factor IX is used to treat Haemophilia B (Christmas disease), a blood-clotting disorder caused by a deficiency of endogenous Factor IX (see also HAEMOPHILIA).

Feedstock: The raw material used for the production of chemicals.

Fermentation: The anaerobic (without oxygen) biological conversion of organic molecules, usually carbohydrates, into alcohol, lactic acid and gases; it is brought about by enzymes either directly or as components of certain bacteria and in yeasts; in general use, the term is sometimes applied to bioprocesses which are not, strictly speaking, fermentation.

Fertilisation: In sexually reproducing organisms, the activation of the development of an egg through the union of sperm with the egg, so combining their genetic complements.

Geep: A chimera; a novel cell fusion animal that is a hybrid between a sheep and a goat; also known as a shoat.

Gene: (Gr. *genos*, descent.) A gene is a section of a nucleic acid molecule in which the sequence of bases encodes the structure of, or is involved in the synthesis of, a protein.

Gene enhancement: The insertion of additional genetic material into an otherwise normal genome in order to enhance a trait perceived of as desirable; an example is the insertion of additional growth hormone genes into the genome of a normal individual in order to increase his or her height or rate of growth over what is considered to be the normal level.

Gene expression: The mechanism whereby the genetic instructions in a given cell are decoded and processed into the final functioning product, usually a protein.

Gene mapping: Determining the relative locations (loci) of genes on chromosomes.

Gene probe: A short piece of DNA or RNA used to detect the presence of complementary sequences in other nucleic acid molecules.

Gene sequencing: Determining the sequence of bases in a molecule of DNA.

Gene splicing: see *IN VITRO* GENETIC RECOMBINATION.

Gene therapy: The correction of the effect of a genetic defect in an organism or cell by direct intervention with the genetic material; one method under investigation is gene replacement therapy in which additional foreign DNA is inserted to compensate for the malfunctioning gene; another approach would be to activate dormant genes within the genome whose function would substitute for the missing function of the malfunctioning gene.

Genetic code: The relationship between the sequence of bases in the nucleic acids of genes and the sequence of amino acids in the proteins that they code for.

Genetic disorder: A disorder which is associated with a specific defect in the hereditary material; may or may not be congenital (for example, late-onset genetic disorders are not manifested at birth) and may or may not be inherited (for example, chromosomal abnormalities which are newly-arisen).

Genetic enfeeblement: see BIOLOGICAL CONTAINMENT.

Genetic engineering: The manipulation of heredity or the hereditary material; the direct and deliberate attempts by humans to influence the course of evolution and to alter its products; the basic techniques are mutagenesis and hybridisation, which introduce genetic variation, and artificial selection which biases quantitatively the genetic variation of subsequent generations; genetic engineering using artificial selection and traditional hybridisation through cross-breeding has been long practised; artificial mutagenesis is a twentieth century development; in the second half of the twentieth century all three techniques have been developed by *in vitro* methods including microgenetic engineering and cell fusion.

Genetic determinism: The theory that the phenotype is an innate and essentially unchangeable expression of the genotype.

Genetic fingerprinting: see DNA FINGERPRINTING.

Genetic manipulation: see MICROGENETIC ENGINEERING.

Genetic marker: An identifiable feature encoded in the genetic material of an organism; an example of the use of genetic markers is the insertion of a small unique and inactive piece of DNA into the genome of organisms to be released into the environment to enable such organisms to be identified upon their recovery from the environment.

Genetic material: DNA, genes and chromosomes which constitute an organism's hereditary material; RNA in certain viruses.

Genetic recombination: The excision and rejoining of DNA molecules; formation of a new association of genes or DNA sequences from different parental origins (see also *IN VITRO* GENETIC RECOMBINATION and RECOMBINANT DNA TECHNIQUES).

Genetic screening: Following Wheale and McNally (1988), a range of techniques used to diagnose phenotypic traits which have or are believed to have a genetic basis; largely used to detect such traits before they become evident, but can also be used to verify the diagnosis of traits after they have become apparent.

Genetically 'crippled': see BIOLOGICAL CONTAINMENT.

Genetics: (L. *genesis*, origin, descent.) That part of biology dealing with both the constancy of inheritance and its variation; the study of the replication, transmission and expression of hereditary information.

Genome: A collective noun for all the genetic information that is typical of a particular organism; every somatic cell in a multicellular organism contains a full genome; the term genome is also applied to the genetic contents characteristic of major groups (for example, the eucaryotic genome) or of a species (for example, the human genome); not all portions of a genome are genes (i.e. genomes include non-coding DNA); genomes do not include the genetic material of extrachromosomal elements, nor of plasmids or viruses harboured by a cell, although this distinction between genomic and non-genomic genetic material is a reflection of a static paradigm of the genome which is increasingly believed to be inaccurate; genomes can be regarded as an ecosystem of genetic elements (see Wheale and McNally (1988) particularly Chapter 4).

Genotype: (Gr. *genos*, race; *typos*, image.) The genetic constitution of an organism with respect to a particular genetic trait, for example, eye colour (cf. PHENOTYPE).

Germ: A popular word for a micro-organism; also used in the eighteenth and nineteenth centuries to describe the hereditary material.

Germ cell: Sex cell or a cell which gives rise to sex cells.

Germ plasm: The term for that part of an organism which passed on hereditary characteristics to the next generation. According to Weismann's germ plasm theory of 1883, the germ plasm was transmitted from generation to generation in germ cells.

Germline: Cells from which sex cells are derived.

Green Revolution: Term used to describe the replacement of traditional crops by high-yield varieties requiring irrigation systems and inputs of fertilisers and pesticides to sustain them.

Growth hormone: In animals, hormone which affects a large number of metabolic processes including the regulation of growth; somatotropin (or somatotrophin) is a growth hormone produced by the anterior lobe of the pituitary gland (see HORMONE).

Haemoglobin: (Gr. *haima*, blood; L. *globus*, a ball.) The red-coloured protein which binds and carries oxygen in red blood cells.

Haemophilia: (Gr. *haima*, blood; *philos*, inclined to.) Constitutional tendency to haemorrhage as a result of abnormal blood-clotting; inherited as a sex-linked recessive single-gene disorder; results in pain, profound anaemia and orthopaedic problems caused by bleeding into the joints; caused by a deficiency of one of the clotting factors of the blood – Factor VIII in the case of haemophilia A (classical haemophilia), and Factor IX in the case of haemophilia B (Christmas disease) (see also FACTOR VIII and FACTOR IX).

Hereditary disease: A disorder of the genetic material which is transmissible from generation to generation.

Hormone: A chemical messenger of the body carried in the bloodstream from the gland which secretes it to a target organ where it has a regulatory effect; an example is insulin.

Host: A cell (microbial, animal or plant) whose metabolism is used for the reproduction of a virus, plasmid or other form of foreign DNA, including vectors and recombinant DNA.

Hybrid: (L. *hybrida*, cross.) A molecule, cell or organism produced by combining the genetic material of genetically dissimilar organisms; traditionally, hybrids were produced by interbreeding whole animals or plants; cell fusion technology and transgenic manipulation are innovations in hybridisation.

Hybridoma: A 'hybrid myeloma'; a cell produced by the fusion of an antibody-producing cell (lymphocyte) with a cancer cell (myeloma).

Hybridoma technology: The technology of fusing antibody-producing cells with tumour cells to produce in hybridomas which proliferate continuously and produce monoclonal antibodies.

Immune system: The body's system of defence against invasion by foreign organisms and certain chemicals.

In utero: (L., *uterus*.) In the womb.

In vitro: (L. *vitrum*, glass.) Literally in glass; biological processes studied and manipulated outside of the living organism.

***In vitro* fertilisation (IVF)**: The fertilisation of an egg cell by sperm on a glass dish.

***In vitro* genetic recombination**: The precise excision and joining of DNA fragments on the laboratory bench exploiting the biochemical tools of the cell – restriction enzymes and DNA ligase – and the inherent pairing affinity of the duplex DNA molecule (see RECOMBINANT DNA TECHNIQUES and GENETIC RECOMBINATION).

In vivo: (L. *vivo*, live.) Within the living organism.

Infection: The invasion of an organism, or part of an organism, by pathogenic micro-organisms, for example, viruses or bacteria.

Innovation: The first introduction of a new product, process or system into the ordinary commercial or social activity.

Insertional mutagenesis: In transgenic manipulation, the mutation of target host cell genes by the integration of foreign genes.

Instar: (L. *instar*, form.) Insect at a particular stage between moults; (insects, which are invertebrates, have a hard exoskeleton (outer skeleton) or cuticle which prevents any increase in the size of the insect except during certain periods of its development when the insect sheds its outer cuticle and increases its volume; this moulting takes place only in the larval or pupal form – adult insects do not grow; each species of insect has a characteristic number of moults/instars).

Insulin: A hormone, the release of which lowers the level of glucose sugar in the blood.

Interferons: A class of proteins released by certain mammalian cells in response to various stimuli, including viral infection, which are thought to inhibit viral replication; undergoing clinical trials as anti-viral and anti-cancer agents.

Intron: A nucleic acid sequence within a gene which is transcribed into RNA but then excised from the RNA transcript before it is translated into protein.

Invention: The first idea, sketch or contrivance of a new product, process or system.

Invertebrates: (L. *in*, not; *vertebra*, joint.) A general term for all animal groups except the vertebrates, i.e., all animals without a backbone; includes insects, worms and crustaceans.

Jumping genes: See MOBILE GENETIC ELEMENTS.

Karyotype: The chromosomal constitution of an individual.

Ketosis: A condition, found in fasting animals and diabetic animals, in which large amounts of ketone bodies appear in the blood (ketonemia) and urine (ketonuria) as a result of the breakdown of body fat.

Lesch–Nyhan syndrome: An X-linked recessive single-gene disorder in which the central nervous system gradually deteriorates; affected children are typically spastic and slow-growing and under compulsion to bite their lips and fingers so that their arms must be restrained to prevent self-mutilation; the biochemical basis is a deficiency of the enzyme hypoxanthine-guanine phosphoribosyl transferase (HPRT).

Ligase: An enzyme which catalyses the joining together of two molecules; for example, DNA ligase catalyses the joining of two DNA molecules.

Mendelian: After Gregor Mendel (1822–84), relating to the Mendelian theory of heredity; the pattern of inheritance exhibited by traits controlled by a single gene in which there is a simple dominant or recessive relationship between alleles.

Microbe: An alternative term for a micro-organism (see MICRO-ORGANISM).

Microbiology: The study of micro-organisms.

Microgenetic engineering: Following Wheale and McNally (1988), the techniques which enable the molecular biologist to decode, compare, construct, mutate, excise, join, transfer and clone specific sequences of DNA, thus directly manipulating the genetic material to produce organisms, cells and subcellular components; applications include scientific research, biotechnology, farming, healthcare and biological defence (see also *IN VITRO* GENETIC RECOMBINATION and RECOMBINANT DNA TECHNIQUES).

Micro-organism: An organism belonging to the categories of viruses, bacteria, fungi, algae or protozoa; micro-organisms – 'animalcules' – were first observed by Anton van Leeuwenhoek (1632–1723) of Delft in Holland in the seventeenth century; until the second half of the nineteenth century, when the nature of their association with putrefaction, fermentation and disease became a focus of microscopy, Leeuwenhoek's discovery of parasites and bacteria remained a 'curiosity', and those who continued to search for *Vermes chaos*, as Linnaeus classified these 'incredibly small animals', were regarded as eccentrics. Molecule: A group of two or more atoms joined together by chemical bonds.

Mobile genetic elements: RNA and DNA sequences that move from one place to another both within and between genomes; popularly known as 'jumping genes'; the family of mobile genetic elements includes viruses, subviral infectious elements, plasmids, RNA introns, messenger RNA molecules, transposable genetic elements and oncogenes; characteristic features include the ability to move pieces of DNA either within or between cellular genomes, the ability to usurp the biochemistry of the host cell to bring about their own replication, the ability to alter the structure of the host cell genome, and the ability to modify the expression of other genetic elements within the host cell; they are believed to play a crucial role in evolution, and are implicated in development and pathology; viruses and plasmids are mobile genetic elements which are used as vectors in gene transfer technology.

Monoclonal antibody: One of a clone of antibodies produced by a hybridoma.

Multifactorial disorder: A disorder in which both genetic and exogenous factors are multiple and interact; the genetic part of multifactorial causation is polygenic (the result of the action of many genes).

Mutagen: (L. *mutare*, to change; Gr. *gennaein*, to generate.) A chemical or physical or other agent which increases the frequency of mutation.

Mutant: An organism or gene which deviates by mutation from the parent organism(s) or gene in one or more characteristics.

Mutation: A change in the genetic material; can refer to changes in a single DNA base pair or in a single gene, and also to changes in chromosome structure and number which are recognisable under the microscope; mutation in the germ-line or sex cells could result in genetic illness or changes of evolutionary significance; somatic cell mutation may be the basis of some cancers and some aspects of ageing.

Neo-Luddite: Person actively opposed to the introduction of new technology; after the Luddites, an organised band of mechanics which went about destroying machinery in the midlands and north of England in the early nineteenth century; derived from Ned Lud, who, in the latter half of the eighteenth century, smashed up machinery belonging to a Leicestershire manufacturer as a protest against mechanisation.

Newcastle disease: Avian influenza.

Nuclear polyhedrosis virus (NPV): Sub-group of the baculoviruses; their name derives from a gene they have which codes for a protein called the polyhedrin protein; during the dispersive phase of their life cycle, many NPVs are encased together in a polyhedral structure comprised of polyhedrin which protects them from adverse environmental conditions and increases their environmental persistence (see also BACULOVIRUS; POLYHEDRIN INCLUSION BODY).

Nucleic acids: Either RNA or DNA; complex organic molecules composed of sequences of nucleotide bases.

Nucleotide: (L. *nucleus*, kernel.) In DNA, a molecular grouping comprised of a base plus a sugar molecule (deoxyribose) plus a phosphate group; DNA is a polynucleotide, that is a molecule comprised of many nucleotides (see BASE).

Nucleus: A region in the cells of eucaryotes, surrounded by a membrane, in which the main chromosomes are sequestered.

Oncogene: (Gr. *onchos*, swelling; L. *genesis*, origin.) Found in every cell of the body, it is postulated that oncogenes are a broad class of regulatory genes that control the activity of other genes; the oncogene theory of cancer is that when an oncogene is activated at an inappropriate stage in the life cycle of an individual, the cell begins to multiply in an uncontrolled way; cellular oncogenes may be derived from viruses which have integrated into the genome of a host cell, or conversely, certain viruses may have acquired cellular oncogenes when they became infectious.

Oncogenic: cancer-causing.

Patent: The exclusive right to a property in an invention; this monopoly on invention gives its owner the legal right of action against anyone exploiting the patented research without the patentor's consent.

Pathogen: (Gr. *pathos*, suffering.) An organism which causes disease.

Pesticides: Herbicides, insecticides and fungicides used in farming and forestry to control the organisms which reduce the quality or quantity of crop yield.

Phage: Abbreviation of bacteriophage, a virus which infects bacteria.

Phenotype: (Gr. *phainein*, to show; *typos*, image.) The manifest expression of the genetic determinants for a particular trait – the genotype (cf. GENOTYPE).

Physical containment: Measures that are designed to prevent or minimise the escape of recombinant organisms.

Plaque: A zone of clearing on the otherwise opaque areas of dense bacterial growth formed when bacteria which have been infected with viruses are grown on nutrient agar; each plaque is derived from a single virus; the plaque assay method exploits this phenomenon to measure the number of bacterial viruses (bacteriophage) suspended in a given volume of medium.

Plasmid: A small circle of DNA usually found in procaryotes, which replicates independently of the main chromosome(s) and can be transferred naturally from one organism to another, even across species boundaries; plasmids and some viruses are used as vectors in transgenic manipulation.

Pleiotropy: (Gr. *pleion*, more; *trope*, turn.) Multiple effects of a single gene, influencing more than one character.

Polyhedrin inclusion body (PIB): Polyhedron-shaped structure made of the crystalline protein polyhedrin containing many nuclear polyhedrosis viruses (NPVs); the dispersive form of NPVs (see NUCLEAR POLYHEDROSIS VIRUS).

Polypeptide: Long folded chains of amino acids; proteins are made from polypeptides.

Procaryotes: Organisms whose genetic material is not sequestered in a well-defined nucleus; includes bacteria and blue-green algae (cf. EUCARYOTES).

Promoter: In transcription, a DNA sequence to which the enzymes which catalyse messenger RNA synthesis bind.

Pronucleus: (L. *pro*, before; *nucleus*, kernel.) Term used for the nucleus of a mature egg cell or mature sperm.

Prophylactic: (Gr. *prophylaktikos*; *pro*, before; *phylax*, guard.) Preventing disease.

Protein: (Gr. *Proteion*, first.) Proteins are polypeptides, i.e. they are made up of amino acids joined together by peptide links; acting as hormones, enzymes and connective and contractile structures, proteins endow cells and organisms with their characteristic properties of shape, metabolic potential, colour and physical capacities.

Proteolysis: (Gr. *Proteion*, first; *lysis*, loosing.) In digestion, the breaking down of dietary proteins into their constituent amino acids by enzymes in the gastrointestinal tract.

Protoplast: Plant cell whose cell wall has been deliberately removed; used in plant cell fusion.

Recessive genetic disorder: (L. *recessus*, withdrawn.) The gene mutations responsible for recessive disorders are usually only harmful to individuals who do not have a corresponding normal gene; except in the
case of sex-linked conditions, a recessive disorder is only expressed when
the gene mutation responsible is inherited from both parents (see
DOMINANT GENETIC DISORDER).

Recombinant bacterium/cell/plasmid/vector/virus etc.: Contains
recombinant DNA (or recombinant RNA).

Recombinant DNA: A hybrid DNA molecule which contains DNA from
two distinct sources (see GENETIC RECOMBINATION).

Recombinant DNA techniques (technology): A type of microgenetic
engineering; the combination of *in vitro* genetic recombination techniques with techniques for the insertion, replication and expression of
recombinant DNA inside living cells.

Restriction enzymes: Bacterial enzymes that cut DNA at specific DNA
sequences; exploited by the microgenetic engineer, for example, in the
execution of the precise excision of DNA fragments for *in vitro* genetic
recombination, and in the analysis of differences between DNA
molecules.

Retroviruses: Viruses which encode their hereditary information in the
nucleic acid RNA; retroviruses are being engineered for use as vectors in
human gene therapy.

Reverse transcription: Synthesis of a single strand of DNA using an
RNA template.

Ribo(se)nucleic acid (RNA): A molecule which resembles DNA in structure; acts as an adjunct in the execution and mediation of the genetic
instructions encoded in DNA, for example, messenger RNA (mRNA) is
transcribed from a single strand of a DNA molecule and is the template
on which amino acids align to form the coded protein; RNA has been
observed to function as an autocatalyst in transcription and possibly
controls translation; RNA may have been the primordal genetic molecule
functioning both as an enzyme and as a self-replicating repository of
genetic information; RNA is the repository of hereditary information for
some viruses (retroviruses).

Scale up: The transition of a process from a laboratory scale to an industrial scale.

Scientism: The belief that the scientific approach is objective and the only
rational way to approach any problem. It is argued that this belief promotes a passive acceptance of techniques and technologies which are scientifically based and thus apolitical.

Sentient: (L. *sentiens*, feeling.) Having the faculty of perception; a sentient
being is one who perceives; refers to all animals which have the capacity to
have conscious experiences such as pleasure or pain.

Sentientism: The moral position that holds that it is wrong to cause pain or
distress to any sentient being, unless it is with their agreement, or unless it
will bring unquestionable benefit to that same individual sentient if he or
she is unable to give informed consent.

Sex cells: Cells which fuse together to form a fertilised egg; in human
beings the male sex cell is the sperm, and the female sex cell is the egg
cell, also known as the ovum (pl. ova) (cf. SOMATIC CELL).

Sex chromosomes: Chromosomes that differ in number or morphology in different sexes and contain genes determining sex type; in human beings the sex chromosomal constitution of a normal female is XX and that of a normal male is XY.

Sex-linked (X-linked) trait: Determined by a gene located on a sex chromosome, usually the X chromosome in humans, and hence the term 'X-linked trait' is sometimes preferred.

Shotgunning (in microgenetic engineering): A technique for breaking up the entire genome of an organism into small pieces and then inserting those pieces into host cells, where they are cloned into a gene library for the organism.

Somatic cell: (Gr. *soma*, body.) Any cell of the body other than germ cells (cf. GERM CELL).

Somatotropin/somatotrophin: (Gr. *soma*, body, *trephein*, to increase.) A protein hormone which is produced by the anterior lobe of the pituitary gland and promotes general body growth.

Speciation: The origin of a new species.

Species: (L. *species*, particular kind.) A unit of biological classification; sexually reproducing organisms are classified as belonging to the same species if they can interbreed and produce fertile offspring; the interpretation of what constitutes a species is controversial; it is especially difficult to apply the concept of species to bacteria which are not subject to reproductive isolation.

Speciesism: Philosophy which allows rights to humans but denies those same rights to other sentient animals.

Sustainable development: Environmental policy which requires the sustainable use of natural resources; each generation shall pass on to the next generation an undiminished aggregate of capital assets, including certain natural assets considered to be inviolable, such as the earth's stock of biological diversity.

Taxonomy: (Gr. *taxis*, arrangement; *nomos*, law.) The method of arrangement or classifying, particularly of living organisms. Taxonomic studies have led to the development of a system of classification which divides all living things into two large groups called kingdoms, each of which is divided into a series of major sub-groups called phyla (sing. phylum). Each phylum is further divided into a series of successively smaller groups known as classes, orders, families, genera (sing. genus) and finally species. There is generally only one kind of organism in a species. In modern taxonomy, an organism is named using the binomial system, under which an organism's name is designated by the genus to which it belongs (generic name), followed by the name of the species (specific name). The name is always written with a capital letter and the specific name with a small letter, for example, the taxonomic name of humans is *Homo sapiens*. The Swedish naturalist Carl Linnaeus introduced the binomial system of naming organisms in 1735.

Tay-Sachs disease: A recessive single-gene disorder caused by a deficiency of an enzyme called hexosaminidase A; results in paralysis, dementia and blindness; death usually occurs before the end of the third year of life; there is no treatment.

Telos: (Gr. end) The nature or purpose of a thing or creature.

Teratogeny: (Gr. *teras*, monster; *genos*, birth.) The formation of monstrous foetuses or births; thalidomide was a classified as a teratogenic drug.

Test-tube babies: Popular term for the resultant offspring of successful IVF, implantation, pregnancy and birth.

Tissue plasminogen activator (TPA): Biological drug for dissolving blood clots; may have therapeutic value in the treatment of heart attack patients.

Toxin: A substance, in some cases produced by disease-causing micro-organisms, that is poisonous to living organisms.

Transcription: (L. *transcribere*, to copy out.) In gene expression, the synthesis of an RNA molecule using a section of one strand of a DNA molecule as a template.

Transduction: (L. *transducere*, to transfer.) In genetics, the transfer of genetic material between cells mediated by an infectious mobile genetic element for example, a virus.

Transformation: (L. *transformare*, to change in shape.) In genetics, the process whereby a piece of foreign 'naked' DNA is taken up from the surrounding medium by a cell into which it integrates and gives that cell new properties.

Transgenic manipulation: A type of microgenetic engineering in which genetic material from one species is inserted into the genome of a different species; an application of recombinant DNA technology.

Transgenic organism: An organism which has been microgenetically engineered so that its genome contains genetic material derived from a different species.

Translation: (L. *translatio*, transferring.) In protein synthesis, the conversion of the base sequence of a messenger RNA molecule (mRNA) into a polypeptide chain of amino acids for the construction of a particular protein.

Translocation (chromosomal): (L. *trans*, across; *locus*, place.) In genetics, the displacement of part or all of one chromosome to another.

Transposable genetic element: A genetic element that moves from site to site within a cellular genome; now believed to be a major feature of all DNA (see MOBILE GENETIC ELEMENT).

Trypanosome: Parasitic micro-organism which causes diseases, for example, sleeping-sickness, in humans and other animals.

Utilitarianism: The philosophy that claims the ultimate good to be the greatest happiness of the greatest number and defines the rightness of actions in terms of their contribution to the general happiness; it follows that no specific moral principle is absolutely certain and necessary, since the relation between actions and their consequences varies with the circumstances.

Vaccine: A substance introduced into an animal's body in order to stimulate the activation of the body's immune system as a precautionary measure against future exposures to a particular pathogenic agent.

Vectors: Vectors are self-replicating entities used as vehicles to transfer foreign genes into living cells and then replicate and possibly also express them; examples are plasmids and viruses.

Vermes chaos: (L. *vermis*, a worm.) Vermin, a noxious animal; animals destructive to game; animals injurious to crops and other possessions; noxious persons, in contempt. (Gr. *chaos*.) That confusion in which matter was supposed to have existed before it was reduced to order; confusion; disorder (see MICRO-ORGANISM).

Vertebrate: (L. *vertebrae*, a joint.) Animal with a backbone; fish, amphibians, reptiles, birds and mammals are vertebrates.

Vertical merger: In economics, the consolidation of two (or more) companies which produce goods or provide services at different stages of the same industry; for example, the merger of a brewing company with a hop supplier.

Viroid: Pieces of 'naked' (without a protein 'coat') infectious RNA.

Virus: A minute infectious agent; a mobile genetic element; composed of nucleic acid (DNA or RNA) wrapped in a protein coat; can survive on its own but in order to replicate it must be inside a living cell; viruses are used as vectors in microgenetic engineering.

Name Index

Subject Index

For Key to abbreviations, see List of Abbreviations

Notes on Contributors

Brian Ager has a degree in Biochemistry and has worked as a medical biochemist and in the scientific and technical branch of the UK Department of Health. In 1984 he was appointed to the post of Secretary of the UK Advisory Committee on Genetic Manipulation (ACGM) which has been very active in preparing guidelines for genetic manipulation. At present he is seconded by the Health and Safety Executive (HSE) to the Organisation for Economic Co-operation and Development (OECD) and the Concertation Unit for Biotechnology in Europe (CUBE). He has been involved in the OECD study on recombinant DNA safety considerations and has been participating in discussions at the European Commission on EC-wide legislation on biotechnology.

Dr David Bishop has a PhD in Biochemistry. He is Director of the Natural Environment Research Council's (NERC) Institute of Virology (IoV) in Oxford. He has lectured at Columbia Medical School and Oxford University. He is a member of numerous scientific committees and has co-authored over 200 scientific papers. He is an authority on the release of genetically engineered organisms into the environment.

Eric Brunner has an MSc in Biochemistry and has worked as a Research Fellow at the Institute of Child Health in London. From 1984–8 he worked at the London Food Commission (LFC) on a variety of issues including heart disease, and fertiliser pollution and health. More recently he has been working on bovine somatotropin (BST) and is author of the London Food Commission report on BST, *Bovine Somatotropin: A Product in Search of a Market* (London: LFC) 1988.

Robert Deakin is a Canadian who studied Engineering and Business at McGill University in Montreal, and has worked for Monsanto for 18 years. For several years he was involved with poultry nutrition technology transfer in South-East Asia and Latin America and more recently with international business developments in agriculture and animal nutrition. He is currently engaged in preparations to market bovine somatotropin (BST) in Europe and Africa.

Dr Michael Fox comes from Derbyshire. He has a veterinary degree from the Royal Veterinary College, London, a PhD in Medicine, and a DSc in Animal Behaviour from the University of London. He has worked as Associate Professor of Psychology at Washington University, and from 1976–87 was Director of the Institute for the Study of Animal Problems, Washington, DC. He was Scientific Director of the Humane Society of the

United States from 1980–7 and is currently Vice-President of the Humane Society, specialising in farm animals and bioethics. He is also Director of the Centre for Respect of Life and the Environment. He is contributing editor to *McCall's Magazine*, runs a syndicated newspaper column called 'Animal Doctor' and has written over 30 books. In the last few years he has written much on the subject of genetic engineering. His video film, *Silent World*, is available from the Humane Society of the United States and from the Athene Trust, England.

Dr Wolfgang Goldhorn studied veterinary medicine in Leipzig and Munich. He worked as a veterinarian in Bavaria for ten years and since 1967 has been a Veterinary Officer. From 1979–81 he worked for the European Commission in Brussels as an animal welfare expert. He is currently Veterinarian Director of the West German State Veterinary Service and is Co-ordinator of the Veterinarians Organisation for Animal Protection. At the Open Hearings on Farm Animal Welfare at the European Parliament in 1986 he presented the case of the sufferings caused to laying hens in the battery cage system.

Benedikt Haerlin MEP was elected as a member of the European Parliament as a non-Party member on the list of the Green Party in 1984. He studied philosophy and psychology at the universities of Tuebingen and Berlin. He is the founder of the International Gen-ethic Network and a member of the European Parliament's Committee on Energy, Research and Technology and the Committee on the Environment.

Alan Holland is a Lecturer in Philosophy at Lancaster University. He is a member of the Society for Applied Philosophy for which he has helped to organise meetings on animal and environmental issues. He is a keen gardener and keeps free-range hens.

Andrew Lees has a degree in Zoology and Botany and has worked for the Nature Conservancy Council (NCC). He joined Friends of the Earth (FoE) in 1986 as their Countryside and Pesticides Campaigner and is now their Water Pollution and Toxics Campaigner. He has appeared many times on UK television to discuss environmental issues.

The Reverend Dr Andrew Linzey is Chaplain and Director of Studies of the Centre for the Study of Theology at Essex University. He has written or edited eight books which include studies of Christianity and the rights of animals, research on embryos, and animal welfare. He is one of the few voices within the church to make a clear case for a recognition of the rights of animals.

Ruth McNally is a Director of Bio-Information (International) Limited. She has a degree in Genetics from Nottingham University and is completing a programme of PhD research on genetic screening. She has written articles on genetic engineering and the new reproductive technologies and has appeared on British and American television. She is co-author of *Genetic Engineering: Catastrophe or Utopia?* (Hemel Hempstead: Wheatsheaf Books; New York: St Martin's Press) (1988).

Dr Caroline Murphy has a degree in Animal Genetics and the Politics of Modern Biology, and a PhD on the History of the Radiotherapy of Cancer. She has worked as a Wellcome post-doctoral Research Fellow at Manchester University and as Research Assistant to the Professor of Modern Genetics at Glasgow University where she was involved in the mapping of human genes. Her team was the first in the world to map the Kappa light-chain immunoglobin gene to chromosome 2. She is at present the Education Officer (Ethics) for the RSPCA.

Peter Roberts was a dairy farmer for several years until he and his wife found their way of farming to be incompatible with their beliefs. They sold their farm and set up the first company in the UK to distribute textured vegetable protein – the vegetarian alternative to meat. His increased concern about factory farming practices led to the founding of Compassion in World Farming (CIWF) to campaign specifically for an end to intensive livestock production. CIWF has grown to more than 10,000 members. In 1986, the Athene Trust was established as the educational wing of CIWF. In 1984 Roberts took a court case against a veal crate farm (in which calves were chained by the neck and unable to turn around in their narrow crates). Although the case was lost, the resulting publicity helped to speed up the end of this method of veal production, which from 1990 is illegal in the UK.

Edward Lee Rogers has a law practice in Washington DC with a focus on environmental and public health issues. He conducts environmental litigation for the Foundation on Economic Trends, and its president, Jeremy Rifkin; the Foundation is the leading US organisation speaking out against irresponsible and unsafe developments in genetic engineering. He was the Principal Deputy Assistant Secretary of the US Army (Civil Works), under the Carter Administration, exercising environmental policy oversight for the Army Corps of Engineers water resource projects and water quality regulatory activities. He has been Assistant Attorney General in the State of Maine's Environmental Protection Division, Counsel to the Natural Resources Council of Maine, and General Counsel to the Environmental Defense Fund. He is a member of the Board of Editors of the Biotechnology Law Report and has published several papers on environmental law, biotechnology regulation and the application of patent law to genetically engineered organisms.

Richard Ryder studied experimental psychology at Cambridge University and Columbia University (New York) and was for many years a clinical psychologist in Oxford. In 1970 he published *Speciesism* and in doing so created a new word for the welfarist's dictionary. Since then he has written many books and articles on animal welfare. He was Chairman of the RSPCA Council from 1977–9 and Chairman of the Liberal Animal Welfare Group from 1980–8. He is currently an RSPCA Council member and programme organiser of the International Fund for Animal Welfare.

Dr Joyce Tait is Senior Lecturer in the Systems Department of the Faculty of Technology at the Open University where she lectures mainly on environmental and technology management. From 1984–7 she was seconded to the Economic and Social Research Council (ESRC) as Co-ordinator of their Research Initiative on Environmental Issues and also ran the secretariat of an international research network on the perception and management of pests and pesticides.

John Webster is Professor of Animal Husbandry in the School of Veterinary Science at Bristol University. He has studied the nutritional and environmental requirements for health and production in cattle for over 20 years. He is a member of the Farm Animal Welfare Council (FAWC) and the author of *Understanding the Dairy Cow* (1987). He has also conducted research into the welfare of veal calves and has developed a humane system of rearing such calves: the Access System in which social, physiological and nutritional needs of the animals are fully met. He has been involved in the recent debate over bovine somatotropin (BST) and remains sceptical about the need to administer it to cows in order to produce more milk.

Dr Peter Wheale is Chairperson of the Biotechnology Business Research Group (BBRG) and a Director of Bio-Information (International) Limited. He has a first degree in Economics and an MSc in the Structure and Organisation of Science and Technology from Manchester University, and a PhD on the cereals industry. He has held lectureships in British and US universities and has published widely. He is co-author of *People, Science and Technology: A Guide to Advanced Industrial Society* (Hemel Hempstead: Wheatsheaf; New Jersey: Barnes & Noble) (1986), and *Genetic Engineering: Catastrophe or Utopia?* (Hemel Hempstead: Wheatsheaf; New York: St Martin's Press) (1988).